The DOORHANGER'S HANDBOOK

The
DOORHANGER'S HANDBOOK

Gary Katz

The Taunton Press

COVER PHOTO: **Roe A. Osborn**

Taunton
BOOKS & VIDEOS

for fellow enthusiasts

Text © 1998 by Gary Katz
Photographs © 1998 by The Taunton Press, Inc.
Illustrations © 1998 by The Taunton Press, Inc.
All rights reserved.

Printed in the United States of America
10 9 8 7 6 5 4 3 2 1

The Taunton Press, Inc.
63 South Main Street
PO Box 5506
Newtown, CT 06470-5506
e-mail: tp@taunton.com

Distributed by Publishers Group West

Library of Congress Cataloging-in-Publication Data
Katz, Gary, 1952-
    The doorhanger's handbook / Gary Katz.
       p. cm.
    Includes index.
    ISBN 1-56158-227-1
    1. Doors—Handbooks, manuals, etc. I. Title.
TH2270.K38 1998
694'.6—dc21   98-23134
                                         CIP

Once again to Lawrence, Tristan, and Robin:
"The anchor of my purest thoughts, the nurse,
The guide, the guardian of my heart and soul."
And to Whitney, for whom words aren't necessary.

ACKNOWLEDGMENTS

Carpentry must be learned. I'm indebted to the Schieffer brothers, Al and Royal, who for years have kindly shared with me their tricks and secrets of hanging doors. And to James Cross, Michael Rutledge, and Kevin Katz, I add my appreciation for their helpful carpentry talent and techniques. There have even been moments when Mark Cross has surprised me with a new technique, as has Arnulfo "Adam" Guerrero: Both have prevented countless mistakes. And to all the unmentioned carpenters I've gleaned from but can't possibly mention, I offer my gratitude.

No author is an island. Without Dean Della Ventura and his incomparable, ongoing, and essential lessons in photography, this book would have been a blur. And my appreciation goes to Mike Mandarano, for making sense and beauty out of my crude drawings. I thank all the staff of The Taunton Press who have helped me in many ways, including Roe Osborn, Kevin Ireton, and Mike Guertin. But I particularly wish to thank Carolyn Mandarano: A fine teacher should always be frightening at first—you succeeded at both.

I thank my brother Larry and his increasing patience.

CONTENTS

INTRODUCTION 1

Chapter 1
CHOOSING A DOOR 2

Hollow-Core Doors 3

Solid-Core Hardboard Doors 7

Special-Order Doors 7

Stile-and-Rail Doors 10

Composite Doors 15

Chapter 2
CHOOSING HARDWARE 18

Hinges 18

Latches 24

Dead Bolts and Locksets 30

Astragals for Pairs of Doors 36

Chapter 3
DOOR JAMBS 38

Interior Jambs 38

Exterior Jambs 51

Chapter 4
PREFIT DOORS 70

Prehung Doors 70

Prefit Bifold Doors 87

Prefit Bipass Doors 92

Mirror Bipass Doors 95

Chapter 5
HANGING DOORS 96

Scribing a Door 97

Cutting the Top and Bottom 104

Preparing the Hinge Stile 107

Preparing the Lock Stile 110

Hanging the Door 119

Chapter 6
ADVANCED DOOR HANGING 126

Hanging a Pair of Doors 126

Installing Mortise Locks 145

Chapter 7
UNCOMMON DOORS & UNUSUAL HARDWARE 158

Sidelights and Latches 158

Arched Doors 164

Dutch Doors 166

Double-Acting Doors 169

Invisible Hinges 176

Site-Built Bifold Doors 179

Pocket Doors 187

Chapter 8
WEATHER-PROOFING FOR EXTERIOR DOORS 194

Sill Covers 194

Thresholds 198

Door Shoes 201

Permanent Installation 204

Weatherstripping the Jamb 208

INDEX 214

INTRODUCTION

I remember my first entry door—vividly. I wish I could forget it. All the tools I owned fit easily into the small rusty cross-box in my pickup truck. For door hanging, I had a circular saw, a belt sander, and three sharp chisels, which were a lot sharper than me.

The first mistake I made was running the hinges out the wrong side of the door. My second mistake was mortising and hinging the door before I cut it to fit the jamb. It wasn't until after I had the door swinging that I found the head was out of level. But I solved that problem by moving the hinges down on the door and trimming the top of the door a little—two, three, maybe four times. Finally it fit pretty good, but then the door had to be planed on the strike side—a lot. The elderly and rigid homeowner had kept a sharp eye on me throughout the entire process, but never said a word, until then.

"I knew it," he mumbled. "Jamb's terrible, huh?" he said.

I grunted in agreement and took the door down for the fifth time. The perspiration dripped down my eyes, and I vowed never to hang another door.

"But you seem to be winning," he added and smiled.

I smiled, too, and kept on hanging doors. Though today I couldn't be more ashamed and embarrassed at my early mistakes, I'll always remember that man believed I was winning. Carpentry is like that. It always feels like learning is losing, only it's not. Sometimes it's just hard work.

Most of us in the construction industry have heard these two axioms: The test of a good rough carpenter is the ability to build a perfect set of stairs. The test of a fine finish carpenter is the ability to hang a perfect pair of doors. Finish carpenters invariably strive for that perfect pair, with hinge gaps, head gaps, and strike gaps straight and even, "so you can just get a nickel between the door and the jamb, all the way around."

For some carpenters, achieving the perfect hang seems impossible: Beginners are often intimidated by the process, and many good carpenters spend too much time and too much effort trimming a little here and a little there. But door hanging doesn't have to be all hard work.

In this book, I describe the methods I use to make door hanging easy and enjoyable. If you've never installed a door before, these methods will free you from worrying, so you can concentrate on mastering your tools. If you've installed doors before and found it a time-consuming struggle, I'm sure you'll find helpful the combination of production methods and custom techniques I've collected from other finish carpenters and from my own painful trial and error.

I've ordered the chapters to follow most general experiences: Choosing a door is discussed in the first chapter, while the second chapter covers hinges and hardware. Chapter 3 deals with the foundation for every door—installing jambs. Installing pre-fit doors—because they're so popular today—follows in Chapter 4. Chapter 5 begins the systematic and step-by-step process for hanging doors from scratch. Chapters 6 and 7 cover advanced techniques for especially challenging jobs. Finally, Chapter 8 covers the essentials of weatherstripping.

If you're just starting out at door hanging, I hope the order and information in this book make the learning process more enjoyable and successful; if you're an experienced doorhanger, I hope this book offers you tips to improve your technique and to make the job easier.

Chapter 1
CHOOSING A DOOR

HOLLOW-CORE DOORS

SOLID-CORE HARDBOARD DOORS

SPECIAL-ORDER DOORS

STILE-AND-RAIL DOORS

COMPOSITE DOORS

This book is about door-installation techniques, but to hang a door you must first buy one. There are so many types of doors available today that buying one can sometimes be more challenging than hanging one. So before I walk you through how to hang a door, I'll discuss the most popular types of doors available today and how they're constructed. This information will help you make an informed buying decision. Choosing hardware can be almost as alarming as choosing a door, so in chapter 2 I'll discuss hinges, latches, and locks. Installation procedures can be found in subsequent chapters.

When I started building homes in the 1970s, it wasn't too difficult to decide which doors to buy. There weren't that many choices. I knew I needed 1⅜-in.-thick doors for the interior and 1¾-in.-thick doors for the exterior. My alternatives were four-panel or six-panel Douglas fir doors and solid-core or hollow-core doors with hardboard skins or veneers in birch, ash, or oak. Cost was certainly an issue, but each type of door would also affect the look and feel of the house, and I was the decorator as well as the contractor.

When I look back and compare the choices available then to the choices available today, it seems like the difference between counting the coins in my pocket and counting the grains of sand on the beach. There are countless types of doors available today. There are even stores devoted exclusively to selling doors. It would be impossible to discuss all of these doors, so I will limit

Inside a Hollow-Core Door

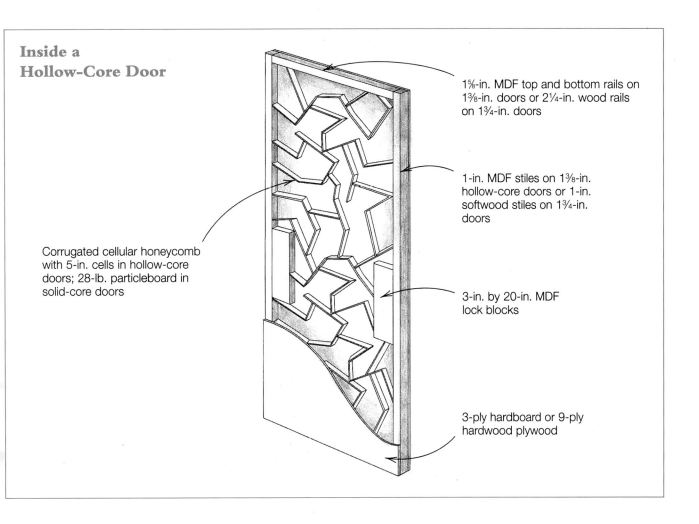

- 1⅝-in. MDF top and bottom rails on 1⅜-in. doors or 2¼-in. wood rails on 1¾-in. doors
- 1-in. MDF stiles on 1⅜-in. hollow-core doors or 1-in. softwood stiles on 1¾-in. doors
- Corrugated cellular honeycomb with 5-in. cells in hollow-core doors; 28-lb. particleboard in solid-core doors
- 3-in. by 20-in. MDF lock blocks
- 3-ply hardboard or 9-ply hardwood plywood

my discussion to prominent types on the market today, beginning with the most common—hollow-core doors.

HOLLOW-CORE DOORS

Hollow-core doors are constructed of two flat skins laminated to a lightweight frame, as shown in the drawing above. Only the edges of the frame are visible because the skins are trimmed flush with the frame, which explains why these doors are also called flush doors. The two long legs of the frame are called stiles, and the two short horizontal members—one at the top of the door and one at the bottom of the door—are called rails. A 3-in. by 20-in. lock block is centered on each stile and provides additional support for hardware. Hollow-core doors aren't really hollow: The center is filled with a corrugated cardboard material. The cardboard and all frame members are glued on edge to the outside skins, forming a strong and resilient lamination.

Hollow-core door construction hasn't changed that much since 1937, when Paine Lumber Company first mass-produced hollow-core doors—at the rate of 3,000 a day. Initially, hollow-core doors did not meet with great approval: The millwork industry was accustomed to established stile-and-rail construction and reluctant to change. Furthermore, buyers were uncomfortable with the newfangled product; there was little belief in the quality of hollow-core doors and whether they would bear up to the burden of everyday, long-term use. At that time the construction industry still suffered from the ravages of the Great Depression, but after World War II, plywood technology and powerful economic growth combined to make the hollow-core door the staple that it is today.

The Secret of Hollow-Core Construction

A typical hollow-core door plant, like American Door Manufacturing, in Stockton, California, warehouses mountains of particleboard door cores, door skins, precut stile-and-rail components, and glue. (Each door requires so much glue that huge tanks of it are needed.) Actual door production begins with the glue-up process. Polyvinyl acetate (PVA), a water-based, environmentally friendly glue, is used by many door manufacturers because it's non-toxic and sets in 30 minutes.

The stiles, rails, and lock blocks are run through a glue spreader, which spreads an even coating of glue on both faces of each piece.

The three-man crew works in sync: one man operates the glue-up machine (top left), a second man feeds the stack two skins at once, and a third man carefully lays in the stiles, rails, blocks, and corrugated backing (top center). When complete, the wet doors are rolled into a press, where they remain for at least an hour (top right). These raw doors are ½ in. over nominal size in width and height, so when the glue has dried, the door is trimmed to exact size.

The first machine trims the width of the door by cutting both edges in a single pass (bottom left). The second machine cuts the top and bottom of each door (bottom right). After being trimmed, the edges are primed and the doors are wrapped with plastic for shipment.

The door-making process begins with the glue-up stage, where a crew works on a growing pile of doors. A hydraulic lift allows the pile to sink into the floor so the work is always at waist level.

One crewman sets each stile, rail, and lock block by hand, while a second crewman waits to cover the frame members with another set of molded panel skins.

When the stack is tall enough it's conveyed to the presses for an hour of high pressure.

After the glue has dried, the doors are sent through the cutting machines. The first machine trims the stiles, so that the door is the right width.

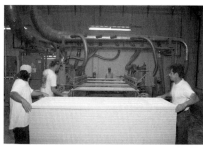

The doors are automatically conveyed through the second machine, which cuts the top and bottom of each door.

Paint-Grade and Stain-Grade Doors

Door skins come in a variety of materials, with hardboard being the predominant product for paint-grade doors. Birch is the usual choice for stain-grade doors, although quality birch skins are now in short supply, so Philippine mahogany, oak, ash, and maple are growing in popularity. The stile and rail components differ greatly between paint-grade and stain-grade doors, and are an important consideration when choosing a door.

Hollow-core hardboard doors were once as common as white paint, and they're still used in a vast number of homes—if not for architectural and stylistic reasons, then because they're the least expensive. The smooth surface of hardboard paints easily and requires little preparation. Because hardboard has a slick surface, it takes any type of finish and is a good base for high-gloss enamel and lacquer. Hardboard is also dense and won't scratch or dent easily. If they aren't abused, hollow-core hardboard doors will last indefinitely.

Hardboard skins (on the left) are predominant on paint-grade doors. Birch (on the right) and other hardwood skins like mahogany, oak, and ash are also common.

Comparing Common Doors

Door type	Paint or stain	Price (approx.)
Hollow-core hardboard	paint	$20
Hollow-core birch	either	$36
Hollow-core oak	stain	$42
Hollow-core molded panel	paint	$36
Solid-core hardboard	paint	$42
Solid-core birch	either	$57
Solid-core oak	stain	$63
Solid-core molded panel	paint	$66
Solid-core combo	paint	$92
Solid-core raised panel (MDF)	paint	$221
Solid-core type 1 glue (waterproof)	—	$46
Solid starved lumber-core	—	$90
Special layout	—	$35 approx. additional cost
Solid-core fire door (20 minute)	—	$52
Stile-and-rail raised panel (MDF)	paint	$152
Stile-and-rail raised panel (Douglas fir)	stain ("C" premium panel)	$204
Stile-and-rail arched door	stain	$500 +
Fiberglass	either (Fiber-Classic)	$170
Steel	paint (with combo window)	$170 (dual glazed)
Clad	paint exterior/ paint or stain interior	depends on manufacturer and design

Notes:
All doors are priced as 3 ft. wide by 6 ft. 8 in. tall.
Hollow-core doors are priced as 1⅜ in. thick.
Solid-core doors are priced as 1¾ in. thick.

Most of the interior framing components (stiles, rails, and lock blocks) for paint-grade hollow-core doors are manufactured from medium-density fiberboard (MDF). MDF is similar in some ways to particleboard, but MDF is formed from finer sawdust fibers mixed with adhesive then formed under extreme pressure. According to many door manufacturers, MDF is the best choice in a paint-grade door because MDF and hardboard swell, contract, and move similarly, which prevents delamination (a problem occasionally found in mixed-material door components). MDF has considerable density and holding strength if fasteners are driven into the face of the material (called the face grain). All door stiles are positioned on edge to ensure that hinge and lock mortises are machined into the face grain. The top and bottom rails are positioned flat, with the grain of the MDF parallel to the door skin, so that bipass hardware can mount on the surface of the door with the screws penetrating the face grain.

The interior components for stain-grade doors, on the other hand, are made from solid wood that sometimes doesn't match the door skin. Many inexpensive stain-grade doors are manufactured with Douglas fir or pine stiles. Blending the stain color between the stiles and the hardwood skins on these doors can be a challenge. For a little more money, stain-grade doors can be purchased with hardwood stiles that match the skin on the face of the door.

Stain-grade doors are manufactured with different types and grades of hardwood skins, too. Common birch skins often contain a wild mixture of dark and light grain colors. For an additional charge, hardwood doors can be ordered with "select" skins that are more closely matched in grain color. Manufacturers also offer skins that are color-matched and book-matched. Color-matched skins, especially in birch, resemble each other in grain density, direction, and color. Book-matched skins are almost photographic duplicates because they're cut from the same section of the same tree.

The Colonist six-panel (right) was the first molded-panel door manufactured. Today more styles have been added, like the Coventry four panel (left).

Hollow-core molded-panel doors are available with arched, or eyebrow-raised, panels, like the four-panel Carmelle (left) and the two-panel Classique (right).

Molded Panels

In the mid-1970s the world of hollow-core doors changed abruptly. Jen-Weld Door Company, in Klamath Falls, Oregon, pioneered a door with embossed hardboard panels that resembled more expensive stile-and-rail, raised-panel doors. The first molded-panel door, which replicated a Douglas fir six-panel door, was named the Colonist (see the left photo on the facing page). That name stuck to all molded-panel doors like chewing gum. Today the term Colonist is synonymous for all molded-panel doors, even though they're now made by numerous companies and many other styles are now available, including the Coventry (see the door on the left in the left photo on the facing page), the Carmelle (see the door on the left in the right photo on the facing page), and the Classique (see the door on the right in the right photo on the facing page). Several of these doors are also available in smooth surfaces, without embossed wood grain.

SOLID-CORE HARDBOARD DOORS

Solid-core hardboard doors are as ubiquitous as hollow-core hardboard doors, and for the same reason—they are inexpensive and they complement almost any design style. Like hollow-core doors, solid-core doors are constructed of two skins laminated to a frame of stiles and rails. Solid-core doors are also referred to as flush doors, but unlike hollow-core doors, which are filled with a honeycomb of corrugated cardboard, most solid-core doors are filled entirely with particleboard, which makes them heavy—around 120 lbs. for a 3-ft. by 8-ft. door.

Every door available with a hollow core is also available with a solid core, including all hardwood skins and all molded-panel styles, like the Colonist. Solid-core doors are mostly 1¾ in. thick, though occasionally I install 1⅜-in. solid-core interior doors to inhibit sound penetration and because some clients prefer the feel of a heavier door.

Solid-core hardboard doors are often used for entry doors and are manufactured with different designs in glass lights or plant-on wood trim. Solid-core rear exit doors with small operable windows are popular. Because manufacturers are able to cut numerous holes into solid-core doors, an endless number of designs are available. One newcomer is a raised-panel door manufactured from a solid-core hardboard door. The raised panels are made from MDF, and the rabbeted molding surrounding each panel resembles the sticking on a more expensive stile-and-rail door. Also, because they are made from solid-core blanks, these doors are less likely than a true stile-and-rail door to crack, cup, check, swell, or shrink—as long as they're carefully finished on each face and all four edges (this rule applies to all doors, interior or exterior).

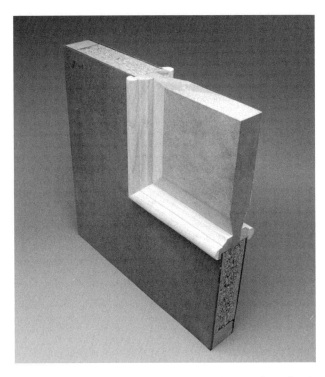

Solid-core hardboard blanks are used to make raised-panel doors, too.

SPECIAL-ORDER DOORS

Standard hollow-core doors come in these widths: 1/3 (1 ft. 3 in.), 1/4, 1/6, 1/8, 1/10, 2/0, 2/2, 2/4, 2/6, 2/8, 2/10, 3/0, 3/6, and 4/0. Standard solid-core doors are manufactured in the same sizes, but beginning with 2/0 doors. However, just because you don't find the door you want at your local lumberyard, don't give up. Solid-core and hollow-core doors can be ordered to suit almost any requirement. Remodels, for instance, often require doors that are not standard sizes, in which case a custom-cut door can be ordered. Some installations, like sliding doors and pivot doors, use special hardware that needs additional backing inside the door, in which case a custom layout door can be ordered. For exterior doors, water-resistant glue is a prerequisite.

Custom Cut

I frequently order custom doors, particularly for doors shorter than 6 ft. 6 in. or more than ½ in. narrower than stock sizes. Because the interior stiles and rails in flush doors are very small, trimming too much off a stock door will compromise its structural integrity and possibly cause early failure. If too much is cut off the bottom of a flush door, the particleboard core might be exposed and the door will rapidly deteriorate. If too much is planed off the stiles of a door, the skins might delaminate, the hinge stile might pull loose from the core, or the latch screws might not hold the lockset firmly in the lock stile. For a minimal additional charge, most manufacturers will supply special custom-cut doors.

Custom Layout

There are circumstances in which a standard size door will suffice but in which the installation includes an unusual type of hardware that requires special backing. Heavy pocket doors often require larger top and bottom rails so that the hardware has more purchase on the top rail and a kerf can be cut in the bottom rail for a floor guide. Pivot hinges and Soss hinges require wider stiles. Door manufacturers offer custom interior component layouts so that stiles, rails, and lock blocks can be sized according to needs (see the drawing below). I often

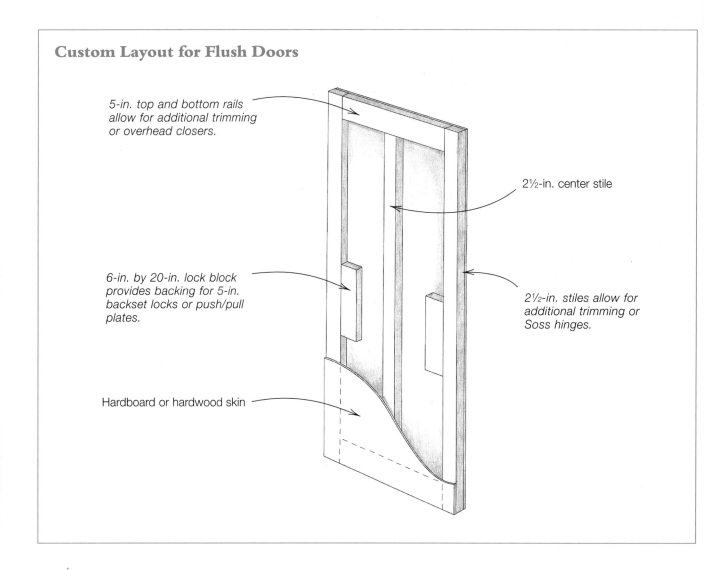

Custom Layout for Flush Doors

- 5-in. top and bottom rails allow for additional trimming or overhead closers.
- 2½-in. center stile
- 6-in. by 20-in. lock block provides backing for 5-in. backset locks or push/pull plates.
- 2½-in. stiles allow for additional trimming or Soss hinges.
- Hardboard or hardwood skin

Door Warrantees

Most door manufacturers offer similar warrantees. It's important to be aware of those policies so that your door installation will meet manufacturers' requirements and be warranted against defects.

Most manufacturers will not guarantee any solid-core door installed in an exterior opening that was made without Type 1 glue. Most manufacturers will not warrantee any solid-core door—with Type 1 glue or not—that's installed in an unprotected exterior opening that has no overhang. Doors manufactured from wood products are dependent upon glue joints. When a wood door is exposed to moisture, it swells; when the door drys, it contracts. The swelling and contraction due to moisture penetration will crack the joints in a door, regardless of the kind of glue used. Once the joint cracks, more moisture will penetrate into the door, and the door will sag and eventually fail. That's why the finish is so important.

It's imperative that all doors are finished completely—at least one coat of primer and two finish coats on all six sides. And to satisfy warrantee require-ments, lap the paint 1/16 in. so that water cannot penetrate between the glass and the wood stop.

Furthermore, most manufacturers will not warrantee exterior doors that are painted black. Black absorbs heat from the sun, which bakes the wood and the glue joints in a door, causing early failure.

These warrantee exceptions may seem picky but they're actually reasonable, especially when compared to a statement I recently found attached to a door: "Any dents, scratches, or imperfections not visible from a distance of 4 ft. will not be considered a product defect under the terms of this warrantee."

---

order 2½-in. stiles for doors that have to be planed irregularly to fit an existing jamb. I order 5-in. top and bottom rails if I'm not sure how much of the door height I'll have to cut. Wider stiles are also essential for Soss hinges (see Chapter 6); larger top and bottom rails are good for bipass closet doors that have to be cut down in order to clear carpeting; and larger, 6-in. by 20-in. lock blocks are necessary for mortise locks.

Some doors require an exceptional amount of hardware. Commercial doors often hold operating hardware such as panic bars, overhead pneumatic closers, magnetic hold-opens, coordinators, and glass vision lights (small windows). The particleboard center of a solid-core door isn't strong enough to support the screws or the direct force of operating hardware. For that reason, manufacturers offer solid-core doors that have staved lumber cores that are made from short, flat pieces of wood glued edge to edge. These doors are extremely heavy and impervious to almost everything, even boys, which is why lumber-core doors are popular among schools.

Water-Resistant Glue

Type 1 water-resistant glue is advisable for all exterior doors. Although this special glue is not waterproof and will not withstand total submersion, a Type 1 glue door will resist abnormal weather conditions and occasional drenching, provided that the door is properly sealed and maintained. A door manufactured with water-resistant glue is slightly more expensive than a door manufactured with standard glue, but the special glue is good insurance against swelling and delamination. Only Type 1 glue doors are warranted for exterior use. Special molded-panel doors for exterior use, like the Elite, manufactured by Jen-Weld, are produced with Type 1 glue.

Fire Doors

Wood fire doors are available with 20-, 45-, 60-, and 90-minute ratings. Each rating describes how long the door should last in a fire. A steel door has to be used when a rating over 90 minutes is required. A standard 1¾-in. solid-core door has a rating of 20 minutes, but it can't be used as a fire door unless it is labeled as such. Every fire door requires a stamped fire label to meet building codes.

Single-family residential construction codes normally require a 20-minute fire door for openings leading directly from a house to a garage. Similarly, multi-family residential codes often require 60-minute fire doors separating residences from 2-hour fire corridors and hallways. All fire doors have to be purchased pre-hung or prefit because installers are no longer allowed to cut them in the field, or they won't meet code. Fire inspectors found instances where fire doors had been planed or cut improperly, which impaired the integrity of the door and the fire rating, especially on doors rated for more than 20 minutes. So not only can fire doors no longer be cut in the field, but they also can't be planed or drilled in the field.

Fire doors rated higher than 20 minutes have a gypsum mineral core, and, to meet code requirements, the wooden stiles are extremely thin. For 90-minute doors the wood is only a veneer. The amount of wood allowed in a door decreases as the rating increases. A composite material—called firestop—is laminated behind the thin wooden stiles and provides the strength necessary to support the door (see the photo below). Firestop is used exclusively for top and bottom rails, lock blocks, and all other interior backing.

STILE-AND-RAIL DOORS

Before the early 1950s, when flush doors came to dominate the market, stile-and-rail doors were the only choice, except for primitive plank doors used on rustic homes, cabins, and barns. The term stile and rail describes the construction technique, as shown in the drawing on the facing page. Two long boards, called stiles, run the vertical height of a door and form a framework through their connection to two or more horizontal members, called rails. The two stiles are normally between 4 in. and 5 in. wide. The top rail is the same width as the stiles, whereas the bottom rail is much wider, between 8 in. and 10 in. Between the stiles, and filling the remainder of the space in the door, are panels. The panels vary in number, size, and shape, from flat to raised and from single to 4, 6, 8, 12, and even more in some custom doors. The joinery connecting the frame varies. Many manufacturers use dowels to strengthen the end-grain joints between the stiles and the rails; smaller shops use dowels, too, though they also rely on mortise-and-tenon joints and loose-tenon joints.

Most stile-and-rail doors also have a secondary molded glue joint, called coping and sticking (see the drawing on p. 12). In cope-and-stick joinery, all interior edges of the stiles and rails are run through a shaper or router equipped with a sticking bit. Sticking profiles vary, though most resemble an ogee-beaded edge. While the

This 60-minute fire-rated door (before veneering) shows the firestop and thin wooden stile separated from the gypsum mineral core and firestop bottom rail.

Stile-and-Rail Door Construction

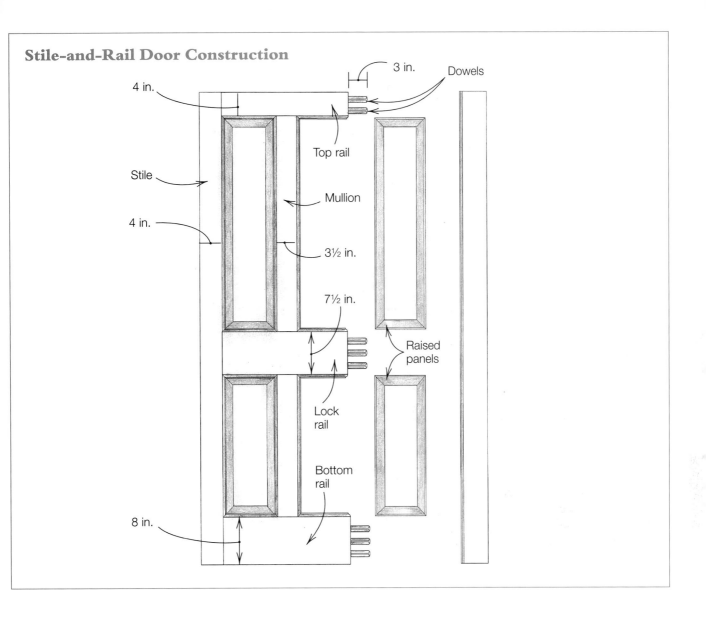

sticking bit cuts a bead on both edges of the stiles and rails, it also cuts a rabbet, or groove, centered in the edge of the stiles and rails. The rabbet is used to secure the center panels of the door.

Coping and sticking the rails to the stiles provides two advantages over simple butt joints: The molded intersection increases the surface area of the glue joint, thus strengthening the joint; and the cope-and-stick joint registers the stiles and rails during glue-up, so that the surfaces of both pieces are almost perfectly flush and require very little sanding when dry. Glue is applied to all the joints in a door, but not in the rabbet where the panels are seated. The panels must float free because the stiles and rails expand and contract seasonally and at a different rate than the panels. I once worked on a home where the painters caulked the raised panels to the door sticking. The painters hoped to eliminate the hairline crack that normally develops between the panel and the sticking, but when the heat came on after the home was finished, the panels cracked and split instead of moving freely.

Coping and Sticking

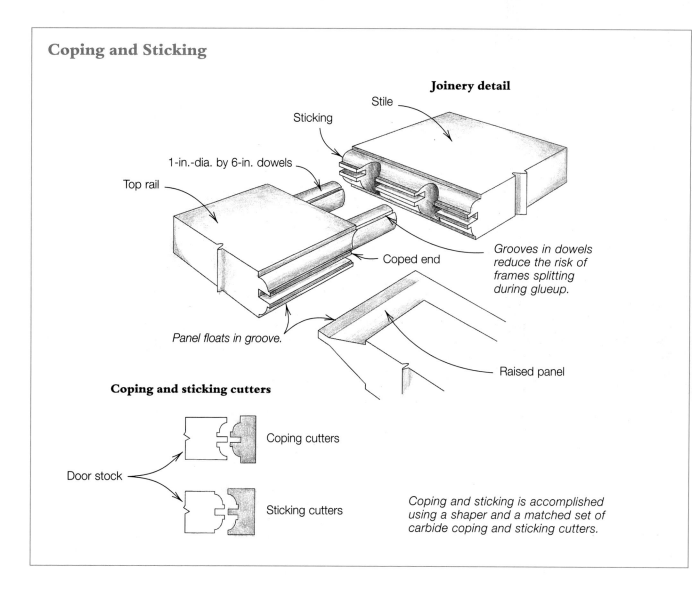

Stile-and-rail doors with flat panels are among the least expensive and are available in several stock designs. Flat panels are also easy to produce in custom layouts. The most common stile-and-rail doors used today have four or six raised panels, though a host of configurations is available, including arched panel doors. Stile-and-rail techniques are also used for the construction of custom doors, like the plank door in the top right photo on the facing page.

Veneered Stiles and Rails
Stile-and-rail doors have always been thought of as solid-wood doors, but today they are rarely made in solid one-piece construction. The materials that comprise modern stile-and-rail doors are as varied as the designs, partly because the price of lumber has risen exponentially since the 1970s and partly because the technology of veneering and laminating wood products has improved substantially. The definition of the phrase *solid wood* has changed, too, as shown in the bottom right photo on the facing page. The term solid wood now applies to doors manufactured from *pieces* of solid wood—scraps, cutoff, and cull—that are glued together then veneered with a thin skin of Douglas fir, oak, mahogany, or other finished material.

Flat-panel and raised-panel paint-grade stile-and-rail doors with MDF panels are available in many designs. On the right is a custom door with two vertical flat panels. On the left is a common six-panel door with raised panels.

Stile and rail techniques are used for almost all custom doors including this arched-panel mission style plank door.

Veneered products are not inferior to true solid-wood one-piece construction. Though I've heard a few complaints of veneers bubbling or blistering, these problems are increasingly rare. On the contrary, there are more problems (cupping, twisting, shrinking, and cracking) in solid-wood products today due more to the growth age of the lumber than to the veneer. Most wood products now depend on second- and third-growth lumber, which doesn't have the stability typically found in old-growth lumber that was once the staple of the millwork industry.

Laminated Stiles and Rails

Small custom shops can't afford the equipment required to glue up and veneer stiles and rails, but small shops are also strained by the double hardship of increasing material costs and declining material quality. To save on lumber costs and to improve the stability of the prod-

Most stile-and-rail doors are no longer made from solid one-piece lumber (left) but are veneered over a substrate of laminated or finger-jointed blocks (right). Paint-grade doors are often made with MDF panels and poplar stiles and rails (center).

CHOOSING A DOOR   13

ucts they make, many small shops build up their 1¾-in.-thick stiles and rails from two pieces of true 1-in. lumber. This material is known as 4/4 because its net size is 1 in. (4 quarters of an inch), as opposed to 1x material (that has a net size of ¾ in.). Laminating stiles from two pieces of 4/4 lumber holds down material costs because full 2-in. lumber is far more expensive. Laminated stiles are also considered stronger than one-piece stiles, which is why they're referred to as *stress-laminated stiles*. Laminating two boards reduces the possibility that the door will twist, crack, cup, or bow.

These two stain-grade doors (pine on the left and oak on the right) are manufactured with veneered stiles, rails, and raised panels.

Finger-Jointed Stiles and Rails

Until recently many paint-grade doors were constructed with poplar stiles and rails or were stiles and rails constructed with a finger-joint core covered by a paint-grade veneer. But in the last couple of years, stiles and rails made from exposed finger-joint lumber have started to appear. While I was skeptical with the early exposed finger-joint stiles, I'm convinced that one day soon finger-joint stiles and rails will be as predominant as MDF panels. The millwork industry, in an attempt to control costs and save trees, is introducing more products made from lumber by-products rather than boards from a tree.

As I've mentioned already, paint-grade doors are available in a wide variety of composite materials, many of which are superior to actual lumber because of stability and cost. In the last few years MDF and MDX (a waterproof version of MDF) have become the most common materials for flat panels and raised panels in paint-grade interior and exterior doors. Door panels made from MDF won't crack, split, cup, check, or bow. MDF machines smoothly and doesn't shrink or develop grain fuzz—and it paints beautifully.

Stain-Grade Stiles and Rails

All door manufacturers produce stain-grade doors using some of the same techniques employed in paint-grade doors. Laminated stiles are frequently seen on stain-grade doors because wood grain and color can be carefully selected to hide the lamination. International Wood Products, one of the largest manufacturers of stain-grade and prefinished entry doors, laminates its stiles and rails from two pieces of solid hardwood.

Stain-grade panels, once made exclusively from solid-wood glued-up boards, are also manufactured in a variety of ways. Generally, solid-wood one-piece panels are still made by small shops that do not have access to expensive technology. Large manufacturers, like Simpson Door Company, have the ability to produce veneered panels and premium split-proof panels, which are formed from alternating laminated layers of substrate and veneered with stain-grade material (see the photo at left).

COMPOSITE DOORS

Not only has the use of solid wood in doors declined, but all lumber and lumber by-products used in doors have declined because of developments made with composite materials. Fiberglass, steel, and aluminum-clad products are now commonly found in the front doors of many homes.

Hollow-core doors were slow to catch on in the Midwest and East Coast, but fiberglass and steel insulated doors were immediately popular in those cold climates, and not only because they keep the cold out. These doors are impervious to weather, and, unlike wood, fiberglass and steel doors last indefinitely.

Fiberglass

In the early 1980s Therma-Tru developed the first fiberglass doors, and a number of companies have since joined the market. Fiberglass doors now capture a large share of the exterior door market.

There are two types of fiberglass doors, but the difference between the two has nothing to do with the number of molded panels or the design. The first type is a true flush door. The fiberglass exterior skin is laminated flush to the stiles, like a hollow-core or solid-core door. Though the edge of the fiberglass skin is visible when looking at the edge of the door stile, most fiberglass skins blend well and almost disappear beside hardwood stiles. These more expensive doors, like Therma-Tru's Classic-Craft (shown in the photo at right), are often manufactured with oak stiles and are meant to replace solid-oak entry doors. The fiberglass skin on a Classic-Craft is also thicker and allows for more detail and depth in the molded panels and grain.

The second type of fiberglass door isn't a true flush door. Though the fiberglass skin is slightly thinner, the real difference between these doors is that the skin is not flush with the stiles but stops ¼ in. short and tucks into a groove cut in the face of the stile. The ¼-in. projection of the stile allows a little room to plane the stiles so that the door will fit an existing opening. However, the stiles shouldn't be planed past the groove. The ¼-in. projection of the stile also creates a different appearance than a standard flush door because when closed, the margins between the door and the jamb seem bigger due to the step in the door skin.

This Classic-Craft door by Therma-Tru is a more expensive fiberglass door than other types due to its finer detail and flush stiles.

Most fiberglass doors on the market have a polyurethane core (shown in the photo on p. 16). Though polyurethane foam degrades over time, it never loses more than 2 points of its R-value, a gauge of insulating capacity. Polyurethane will always have a higher R-value than less-expensive polystyrene (disposable beverage coolers are made from polystyrene).

Both types of fiberglass doors can be painted or stained—or even restained. I watched one demonstration where a Therma-Tru representative stained a fiberglass door a dark oak color. After the door dried, he used 100% mineral spirits to strip the door back to its original condition, then he finished it again, but with a light oak stain. I was impressed. It's hard work, if not nearly impossible, to restain a dark wood door in a lighter color.

CHOOSING A DOOR 15

Steel and fiberglass doors are constructed similarly. The interior voids are filled with either polystyrene foam (center, steel door) or polyurethane foam (left, steel door and right, fiberglass door). Both of these steel doors have thermal breaks.

Most manufacturers supply finishing kits for their fiberglass doors. Special coatings or preparation are not required. Manufacturers recommend any good primer and an oil-based or 100% acrylic latex top coat. If the door is stained, an ultraviolet inhibitor is required in the top coat.

French doors are one of the most popular and practical applications for fiberglass doors. Because most fiberglass French doors are glazed before the core is filled, the polyurethane foam core adheres to the glass and permanently seals the glass to the door, which eliminates water penetration, the primary cause of failure with wood French doors. If the glass breaks the door has to be replaced. These doors are not designed to be cut very much in the field, so attention should be paid to door height when ordering replacement doors for existing openings. When I use fiberglass doors to replace existing wood French doors, I order the doors sized especially for the openings.

Steel

Steel doors are manufactured much the same as fiberglass doors. The outer skin is actually an envelope that encases and adheres to the polyurethane or polystyrene foam core, as shown in the photo above. Steel doors that don't have wooden stiles should have a thermal break (see the drawing on the facing page) so that the cold and moist exterior air doesn't travel through the door to the inside, which can cause frosting on the interior of the door. Many steel doors have wooden stiles that allow some planing and fitting. Therma-Tru, Stanley, Jen-Weld, and Weather Shield all market steel insulated doors in a variety of designs, from French doors to molded-panel entry units. I frequently install steel doors in exterior openings that exit garages, bathrooms, and service entrances, because they last longer than wood doors and are less expensive than fiberglass.

Like most products, steel doors are available in a variety of price ranges. Unfortunately, inexpensive models can be ugly and they often dent easily, which has damaged the reputation of the entire line. Also, some lower-priced steel doors can't be mortised for hinges, while others come with hinge mortises that run out both faces of the door, for right-hand and left-hand applications—a look that's seldom acceptable for an entry door.

The Signature Series steel entry door, manufactured by Weather Shield, is an exception. This door is manufactured with a 1-in.-thick polyurethane core wrapped in a steel envelope. The steel is in turn wrapped with a molded oak veneer, which stains exactly like solid wood. Signature Series doors are expensive but they look just like solid oak doors, and they don't have the maintenance problems.

Aluminum and Vinyl Clad

At the higher end of the price spectrum, along with the Classic-Craft and the Signature Series doors, are clad doors. Clad doors are available from many suppliers, including Anderson, Pella, Kolbe & Kolbe, Weather Shield, and Marvin. Some suppliers use aluminum cladding and others use vinyl. Many people in the industry define fiberglass doors and steel doors as clad doors, too, because all of these doors share similar stile seams: The exterior cladding material is wrapped around the edge of the door and crimped into the stile (similar to steel doors); or the stile is rabbeted or dadoed to accept the cladding. For exterior doors that would otherwise be painted on the outside and stained on the inside, a wood-clad door is a good choice in cold climates.

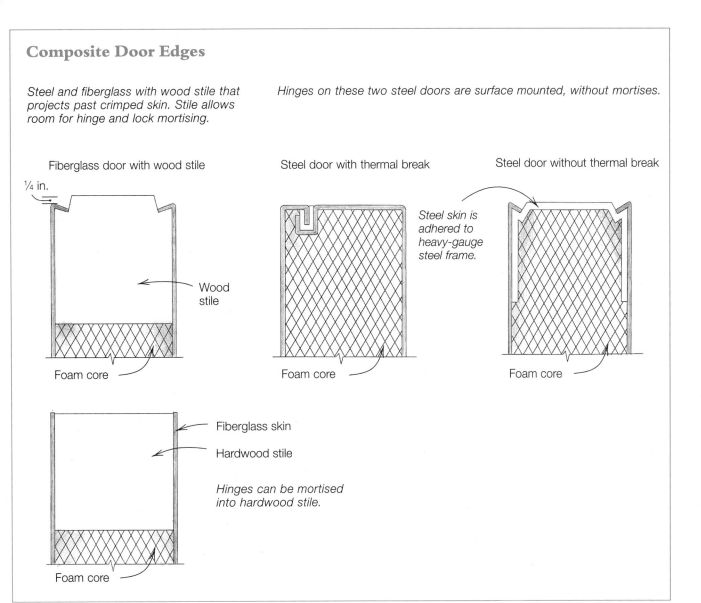

Composite Door Edges

Steel and fiberglass with wood stile that projects past crimped skin. Stile allows room for hinge and lock mortising.

Hinges on these two steel doors are surface mounted, without mortises.

Fiberglass door with wood stile
¼ in.
Wood stile
Foam core

Steel door with thermal break
Steel skin is adhered to heavy-gauge steel frame.
Foam core

Steel door without thermal break
Foam core

Fiberglass skin
Hardwood stile
Hinges can be mortised into hardwood stile.
Foam core

Aluminum-clad doors are available in more colors than vinyl-clad doors, though vinyl cladding has several advantages over aluminum cladding. Vinyl doesn't dent and doesn't require crimping at corners or butt joints where stiles and rails meet. Vinyl cladding is fusion-welded and never separates. Vinyl absorbs the seasonal swelling and contraction of the inner wood core and doesn't rust or corrode, one reason it's preferred on homes near the ocean.

Most clad door manufacturers market prefit units, which include the jamb, doors, hinges, and weatherstripping, but many companies will sell a custom door for an existing opening, too. I have ordered Pella doors and Weather Shield doors for existing wood jambs. In such cases I specify the exact size of the door—width, height, bevel, and swing—and provide the precise hinge layout and lock location. Ordering a custom clad door is tricky, but once a clad door is installed it never requires exterior maintenance.

Chapter 2
CHOOSING HARDWARE

HINGES

LATCHES

DEAD BOLTS AND LOCKSETS

ASTRAGALS FOR PAIRS OF DOORS

Deciding which hardware to use isn't any easier than choosing a door. There are an endless number of products available, and new products are being introduced almost weekly. Without some familiarity, even selecting hinges can be a confusing task. It would be impossible to cover the subject of hardware completely in one chapter, but I'll discuss the most common types of door hardware, including hinges, latches, locks, and astragals.

HINGES

Door hinges, also referred to as butt hinges, are available in every imaginable finish color and in a variety of sizes and shapes. The most important thing to understand about hinges is that different size hinges are designed for different size doors. The relationship between hinge width and door thickness is important because the thicker the door, the wider the hinge has to be for the door to swing clear of the casing, as shown in the drawing on the facing page.

For butt hinges on residential doors, it's safe to follow this rule: 1¾-in. doors use 4-in. by 4-in. (4-in.) hinges; 1⅜-in. doors use 3½-in. by 3½-in. (3½-in.) hinges. If the casing or base is more than 1 in. thick, this rule doesn't apply. I'll explain why.

A 4-in. hinge will work on a 1¾-in. door as long as nothing—no casing, baseboard, or door-trim detail—projects more than 1 in. beyond the jamb. If any trim projects more than 1 in. from the jamb, then the door won't open 180 degrees and lie flat against the wall. In-

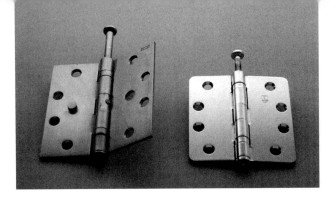

Ball-bearing hinges should be used on all solid wood doors larger than 3/0 x 6/8. Non-removable pin (NRP) hinges (left) in conjunction with a security stud are used on doors that swing out.

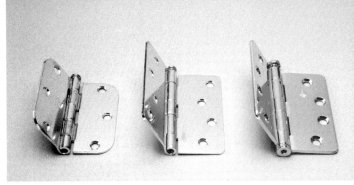

On prehung doors, ⅝-in.-radius corner hinges (left) are commonly used. The two ¼-in.-radius corner hinges (on the right) are popular among carpenters with standard hinge templates. The two hinges on the left are used on hollow-core doors; the thicker hinge on the far right is recommended for solid-core 1¾-in. doors.

stead the door will pinch on the trim and damage the face of the trim. If the door pinches on the trim, it can also damage the jamb, bend the hinges, or even strip the hinge screws. Wider hinges are a must for thick trim. Both 3½-in.- and 4-in.-tall hinges are available in 4½-in., 5-in., and 6-in. widths. Wider hinges, also known as wide-throw hinges, are expensive, but sometimes they're the only answer. If the casing or baseboard you're using is especially thick, or if the door will be installed in a recessed opening, use the formula in the drawing below to determine the necessary hinge width.

A 4-in. by 5-in. hinge is also called a wide-throw hinge because it allows a door to swing wide of thick casing or baseboard.

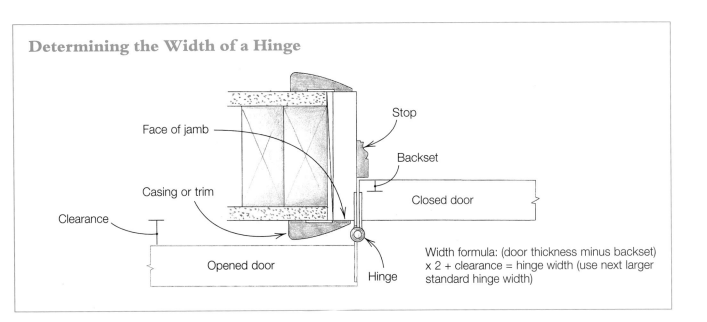

Determining the Width of a Hinge

Width formula: (door thickness minus backset) x 2 + clearance = hinge width (use next larger standard hinge width)

CHOOSING HARDWARE 19

Hardware Finish

Denoting the finish color on hardware is confusing because two systems are used in the industry to identify colors and color tones on metallic surfaces. One method is the U.S. system, which Hagar Hinge Company and Stanley Hardware use. The other is the ANSI/BHMA (American National Standards Institute/Building Hardware Manufacturers Association) system, used by most lock manufacturers. The chart at right simplifies the confusion (an asterisk denotes the most common finish colors).

U.S./Hager	ANSI/BHMA	Finish description
U.S. 3	605	Bright Brass*
U.S. 4	606	Satin Brass*
U.S. 5	609	Satin Brass Oxidized*
U.S. 10	612	Satin Bronze
U.S. 10A	614	Antique Bronze Lacquered
U.S. 10B	613	Antique Bronze Oiled*
U.S. 14B	NA	Bright Black Nickel
U.S. 15	619	Satin Nickel
U.S. 15A	620	Satin Nickel Oxidized
U.S. 17A	621	Stain Black Nickel
U.S. 26	625	Bright Chrome*
U.S. 26D	626	Satin Chrome*
U.S. 32D	630	Satin Stainless Steel*
USP	600	Prime Coat, Beige*
LS	689	Luma Sheen
L1	693	Luma Color Black*
L2	695	Luma Color Dark Bronze
L3	694	Luma Color Medium Bronze*

Round- and Square-Corner Hinges

The most common hinges used today have ¼-in.-radius round corners. Round-corner hinges with a ⅝-in. radius are also available, but they are used primarily by manufacturers of prehung doors and are referred to as prehung hinges. The combination of router bit, template guide, and template used by most carpenters for mortising hinges produces a mortise with ¼-in. radius corners, which explains why these hinges are so common: The hinge fits right into the mortise without additional fuss. Occasionally the design of a home requires square-corner hinges. In any case, hinges larger than 4 in. are only manufactured with square corners—all steel doors and doors in metal jambs use square-corner hinges.

Light- and Heavy-Duty Hinges

Hinges not only come in different sizes and shapes, but they also come in different thicknesses, pin sizes, and knuckle configurations—all of which depend upon the size of the door, the weight of the door, and, surprisingly, how frequently the door is opened. Residential interior doors are rated low frequency because they're opened fewer than 25 times a day. Residential entry doors are rated average frequency because they're opened approximately 40 times a day. Commercial doors, operated anywhere from 400 to 5,000 times a day, are rated high frequency.

Most hinge standards and hinge designations are set by the two largest hinge manufacturers, Stanley and Hagar. Both companies manufacture light-duty and heavy-duty hinges that are suitable for residential use. For hollow-core interior doors 1700 series light-duty hinges are commonly used. A 3½-in. 1741 hinge is 0.085 in. thick; a 4-in. 1741 hinge is 0.097 in. thick (the center and left hinges in the top right photo on p. 19). For heavier solid-core doors, and especially doors over 6 ft. 8 in., the heavy-duty 1279 hinge is advisable because it's thicker: The 1279 3½-in. hinge is 0.119 in. thick, and a 4-in. 1279 hinge is 0.129 in. thick (the right hinge in the top right photo on p. 19).

Doors over 3 ft. wide, heavy entry doors (like oak and maple), and doors installed with automatic closers should have ball-bearing hinges. Ball-bearing hinges won't squeak or squeal under the weight of a heavy door and won't leave tell-tale metal grindings or black metal dust on the hinge knuckles. Ball-bearing hinges are manufactured by many companies, but it's best to use the type in which the bearings are permanently joined to the hinge knuckles, such as the Hagar and Stanley models. If the bearings are not joined to the hinge knuckles, then they fall out when the hinge pin is pulled. It's hard enough to replace the top pin on an 8-ft. door, but if you have to fuss with three or four bearings and climb up and down a ladder each time a bearing falls, it's really frustrating.

Template Hinges

Most residential doors are installed with non-template hinges, which have a staggered screw pattern. The staggered screw pattern helps prevent wooden door stiles from splitting (the stress on the wood grain is eased because the four screws are not in a direct line). But non-template hinges are not reversible—the screw pattern changes from a right-hand door to a left-hand door, so they can't be used on a steel jamb or door. Instead, template hinges are used on steel jambs and doors so that the jambs, doors, hinges, and hand of the door can be reversed.

Many steel jamb legs can be turned upside down to reverse the swing of the door (change the door and jamb from a left-hand to a right-hand opening; see Chapter 5 for more on this), and most steel doors are often reversible, too—some can be turned upside down to change the hand, and some steel doors can be hinged out the front or out the back of the door to change the hand of the door. The screw pattern on a template hinge remains the same regardless of the direction of the hinge—or the hand of the door—so the holes in the hinge always line up with the premachined threading in the jamb and door.

Non-Removable Pins

Hinges with non-removable pins (NRP) should be used on all doors that swing out. Without NRP hinges, the hinge pins could be removed and the door pried out of the jamb. At least that's the theory. In reality, swag—the offset in each hinge leaf designed to reduce the gap between the hinge leaves and the gap between the door and the jamb (see the drawing below)—makes it extremely difficult to remove a door from a jamb with the lock engaged. However, if a door is installed loose in a frame, with large margins or gaps on the hinge and lock stiles, or if a large enough prybar is used, it is possible to pull the pins and pry an exterior swing-out door from its frame. NRP hinges eliminate that possibility, but they aren't the only safety precaution for swing-out exterior doors. Security studs offer similar protection because they interlock with a corresponding hole in the opposite hinge leaf, so that when the doors are closed, even if the non-removable pins are defeated, the hinges cannot be separated.

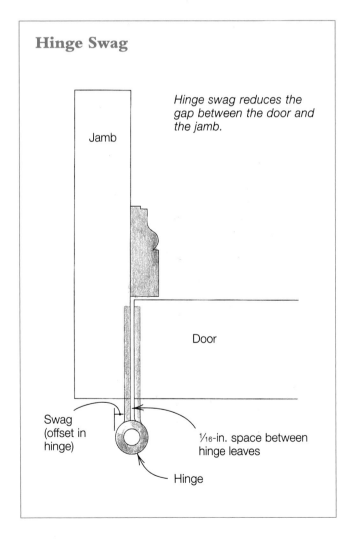

Hinge Swag

Hinge swag reduces the gap between the door and the jamb.

Jamb

Door

Swag (offset in hinge)

1/16-in. space between hinge leaves

Hinge

CHOOSING HARDWARE

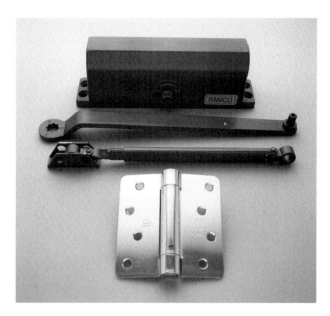

Compared to overhead door closers (top), spring hinges (bottom) are inexpensive, easy to install, and aren't an eyesore, but they offer no speed control and often allow a door to slam shut.

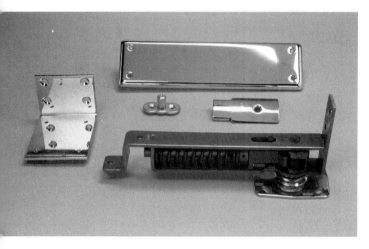

Double-acting hinges allow doors to swing to the left and the right and are usually installed on doors that separate a dining room from a kitchen.

Spring Hinges and Double-Acting Hinges

Spring hinges are used on residential doors that have to close automatically, like the fire-rated door between a garage and a house. I recommend using two spring hinges on fire-rated doors—one on the top of the door and one on the bottom. If only one hinge is installed, it generally has to be wound too tight, and even then rarely closes the door completely. I use a ball-bearing hinge in the center so there will be less drag on the door. Unfortunately, there isn't a wide area of adjustment in spring hinges, and sometimes it seems only one speed works—slam. Overhead hydraulic closers are a good alternative, but many homeowners dislike the large devices (see the photo at left).

Double-acting hinges, like the one in the photo below (more parts are included but couldn't be shown in the photograph), are not door closers. Double-acting hinges are used on doors that swing in both directions, like the door between a kitchen and a dining room. Double-acting hinges are far more expensive than spring hinges and require more time to install, but they can be very appealing (see Chapter 7). Think of all those trips from the dining table to the kitchen sink with both hands full of dishes. That's a good time to have a door that swings open in both directions. And it's nice that the door closes behind you automatically—the mess in the kitchen is never visible to guests at the dining table (and the occasional sound of broken glass is more discreet!).

Soss Invisible Hinges and Piano Hinges

Soss hinges might seem esoteric but they're the best choice for situations where the hinges must be invisible. Mirrored doors in mirrored walls, doors hidden in library paneling, doors in book-matched paneling, and even flush doors in flat white walls can be hidden almost entirely if they're swung on Soss hinges.

Soss hinges are constructed from plated steel and zinc-alloy castings. All the models recommended for use on doors are manufactured with an anti-friction bearing material so the hinges operate smoothly. Hinge 216 is designed for 1 3/8-in. doors, hinge 218 is designed for 1 3/4-in. doors, and hinge 220 is designed for 2-in. doors. Two hinges are recommended for light hollow-core interior doors, and up to four hinges are advised for

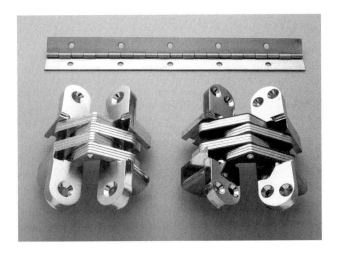

Piano hinges (top) are occasionally used for hidden doors and mirrored doors, but Soss hinges (bottom) work better and are entirely invisible. The Soss hinge at left is for 1⅜-in. doors; the hinge on the right is for 1¾-in. doors.

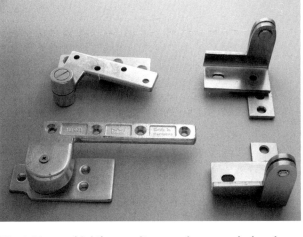

Pivot hinges (right) are often used on wardrobe doors that reach from the floor to the ceiling. They allow doors to open 180 degrees. Floor pivots (left) are an alternative to pivot hinges, but upper floor hinges require a head jamb.

1¾-in. and 2-in. solid-core doors. The spacing on Soss hinges is unusual, at least by today's standards: A few centuries ago, center hinges were often installed closer to the top hinge; likewise, the center Soss hinge should be placed half the distance between the center of the door and the top hinge, to assist the top hinge in carrying the lateral weight of the door (for more on Soss hinge installation, see Chapter 7). Soss hinges are mechanical marvels, and once a door is properly installed on a Soss hinge it will last indefinitely.

Piano hinges are often used when the hinge needs to be hidden, but piano hinges are still visible and never achieve the result of a truly hidden door. Piano hinges are less expensive than Soss hinges, can be purchased in many different lengths, even up to 8 ft., and are easier to install than Soss hinges. But it's troublesome to adjust the fit of a door swinging on one long piano hinge: The piano hinge is a single self-enclosed unit, and therefore it's difficult to make fine adjustments in the fit of the door. Individual Soss hinges can be adjusted so that the door fits in the opening perfectly.

Pivot Hinges

Pivot hinges are often used on specially designed closet wardrobe doors that reach from the floor to the ceiling, without a header. The Stanley 327 heavyweight pivot hinge is the hinge I use most often because it supports doors up to 150 lbs. These hinges are meant to be used on doors that have no jambs. Instead of mounting to a jamb, Stanley pivot hinges require backing at the top and bottom of the door. If the design calls for multiple pairs of wardrobe openings, then posts or blocks have to be installed between each set of doors. If the posts or blocks are objectionable, floor pivots are an alternative.

Floor pivots mount directly to the finished floor surface. A built-up pad is required on carpeted floors. For typical installations a head jamb is required for the top pivot, which, unlike Stanley #327 pivot hinges, requires a break between the top of the door and the ceiling. With a little extra effort a wood pad can be installed in the ceiling, almost flush with the finished plaster or drywall, allowing the top pivot to be buried partially in the plaster and the door installed right to the ceiling.

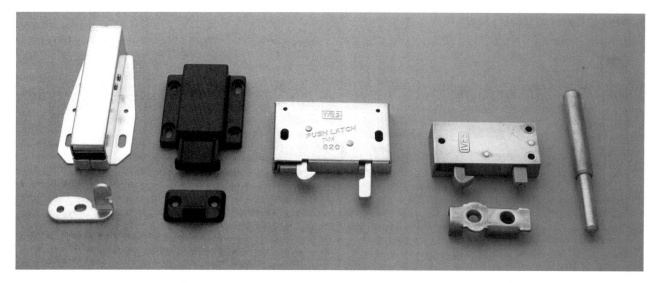

Plastic and light-duty metal touch latches (the first three from the left) are meant for cabinet doors, not for full-size doors. Ives manufactures three sizes—823, 824, and 825 (825 is shown second from the right)—in the same design, each for different size doors. Auxiliary pushers apply a little more pressure to flush doors and make recessed doors easier to open (far right). The Ives 821 light-duty latch is on the far left.

LATCHES

Every swinging door not only needs a hinge but also a latch, so that it won't swing all the time. Latches are simple mechanical devices that operate from only one side of a door. There are many different types of latches, but only a handful are used commonly. Having just discussed invisible hinges for hidden doors, I'll begin with touch latches, because they're often installed with invisible hinges. Afterwards I'll cover ball and roller catches for doors with dummy handles, casement latches for sidelights, slide bolts and flush bolts for stationary doors, and door stops.

My primary concern when choosing hinges, latches, and locks is durability. If the product doesn't hold up my reputation is hurt as well as my profit margin—call-backs are expensive. When it comes to selecting hardware, the old axiom holds true: You get what you pay for. Much of the hardware I'll tell you about is duplicated by less-expensive, imported products that may better fit your budget, though I've seldom found it reasonable to trim the rough edges of a budget by sacrificing durable hardware.

Doors installed with concealed hinges like Soss hinges and piano hinges often require touch latches. Because there's no hardware visible on the face of the door, a pressure-sensitive touch latch is used to secure the door while it's closed flush with the wall. Pushing slightly on the face of the door releases the latch. The plunger on a touch latch helps push the door proud of the jamb so that the edge of the door can be grasped and opened. There are several touch latches available. Most of the light plastic latches and magnetic touch latches work well on cabinet doors. Some plastic touch latches work well on small doors and have the added advantage of being easy to install, but plastic and magnetic touch latches aren't heavy enough or strong enough for full-size doors.

H. B. Ives manufactures a full line of touch latches in the same box design, each meant for a different size door. These latches work well: They have the strength to latch and hold a door closed, as well as the power to unlatch and push a door open. I've found the Ives 825 (second from the right in the photo above) to be the most useful all-around touch latch; even for 2/0 x 6/8 hollow-core doors it's not too strong.

Backset

Backset is the distance a lock is centered from the front edge of a door, as shown below. Locks in commercial buildings and on doors with very wide stiles often have a 2¾-in. backset, so that the lock is farther from the edge of the door and centered more in the door's lock stile. Residential locks generally have a 2⅜-in. backset because stiles in residential doors once commonly made 4½ in. wide are now only 4 in. wide. One nice advance in lock engineering and a pleasant advantage to tubular latches is the adjustable backset latch, which is now commonplace from most tubular lock manufacturers. Cylindrical latches, like Schlage's A series, are not adjustable. The wrong backset latch means one more trip to the store. (For more on cylindrical and tubular locksets, see pp. 30-31.)

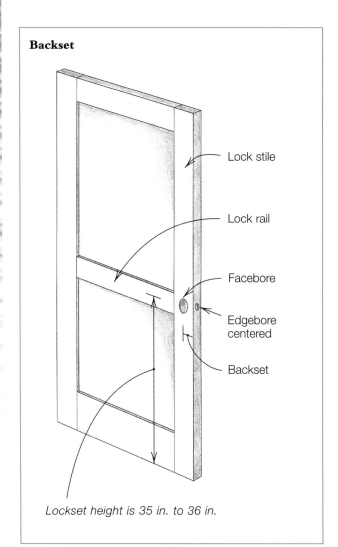

Backset

- Lock stile
- Lock rail
- Facebore
- Edgebore centered
- Backset

Lockset height is 35 in. to 36 in.

The Schlage D series commercial lever set (foreground) is a cylindrical lock because the latch interlocks with the cylinder, whereas the Kwikset lock (upper left) is tubular because the tailpiece at the back of the handle penetrates the center of the latch. The Baldwin Images lockset (upper right) is also tubular and the backset is adjustable.

CHOOSING HARDWARE

Roller catches (far left) are the least expensive choice of all latches, and they're easy to install. Ball catches (second from the left) take a little longer to install because they aren't exposed on the jamb. Roller latches are the best: They're adjustable and work well on any size door (Ives 335, second from the right, with T-strike and D-strike). The Quality 93 (far right) has dual springs and exposed adjustment screws.

Quality casement latches (Ives 66 on the right) are more durable than smaller imports (left). Several strike designs are available, like rim strikes for swing-in openings (left) and mortise strikes for openings that swing out (right).

Ball Catches and Roller Catches

If pulls or knobs are installed on the face of the door, then touch latches aren't necessary and ball catches or roller catches are the best choice to hold the door closed. Inexpensive and easy to install, surface-mounted roller catches, like the Ives 338 (far left in the photo above), are often used on closet doors. Ball, or bullet, catches are the next step up in quality and price. Ball catches are installed in the top of a door, and a matching strike is installed in the head jamb. Ball catches will engage from any angle, which makes them useful on bi-fold doors (see Chapter 7), and they are adjustable. I frequently install Ives 347 ball catches (second from the left in the photo above), which have two adjustments, one for door clearance and one for holding power.

For heavy solid-core doors or doors over 6/8 high, ball catches rarely have the holding power, so the best answer is a roller latch. These latches are powerful and adjustable, so they can be used on any door. To make adjustments, most roller latches have to be removed from the top of the door, but the Quality 93 roller latch (far right in the photo above) has two adjustment screws

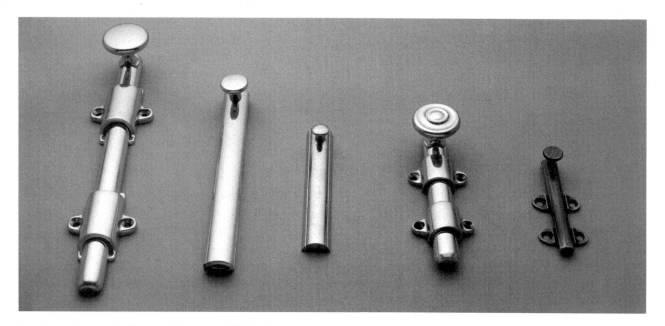

Small surface bolts like the 4-in. Ives 043 (far right) are popular on windows. Larger slide bolts are common on doors, like the Ives 144 (second from the left), available in lengths from 6 in. to 12 in. Decorative surface bolts, like the Ives 8-in. 353 (far left) are available in lengths from 8 in. to 24 in. Dutch-door bolts are used exclusively on Dutch doors (second from the right). In the center is a standard 6-in. slide bolt.

mounted on the surface of the latch. Hagar manufactures a similar roller latch, the 1442.

Casement Latches

I've always liked latches because they're so easy to open and close. Casement latches are sleek and effective yet simple, and they look fine. Casement latches lend weight and feeling to hinged windows and sidelights, like the leather belt and brass buckle that secured the hood on an old 1955 MGA sports car (casement latches are more dependable, though). I prefer the larger, heavier casement latches (on the right in the bottom photo on the facing page) to the smaller imported brands (on the left in the bottom photo on the facing page). They have a solid and secure feel when they're locked, and they don't wiggle.

Slide Bolts and Surface Bolts

Slide bolts are different from surface bolts (see the photo above). Slide bolts are manufactured from two pieces of metal, a bolt and a base, and the bolt slides within the base. Surface bolts are usually made with three or more pieces—the bolt and a minimum of two separate base pieces. Whether you're buying a slide bolt or a surface bolt, it should have a heavy and solid yet lubricated feel as it slides in its base. Higher-quality bolts are retained by and ride on bearings trapped between the base and the bottom of the bolt. Less expensive models are retained by thin springs and the friction of the bolt trapped within the folds of the base. Higher-quality bolts are well designed and engage their strikes easily. They also have fewer installation problems. Poorer-quality imitations often aren't engineered properly, and the bolts fail to mate with the strikes. The price of the hardware is usually a good indication of the quality: A little more money probably means you'll get a product that works better.

Large surface bolts have large knobs, as opposed to the fingernail-size knobs on slide bolts. These long surface bolts work great on tall doors. On an 8-ft. door, a long surface bolt is easier to reach, and the large knob allows plenty of leverage. Even if the top of the door is warped a little away from the door stop (a common problem on tall doors), the force of throwing the bolt will close the door tightly against the jamb.

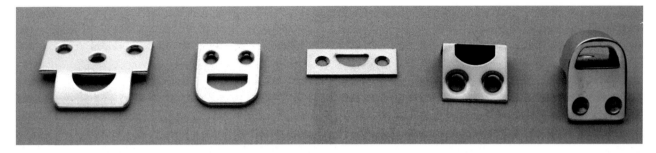

Strikes for surface bolts vary widely. On the left are two universal strikes in different designs. In the center is a mortise strike. Both strikes on the right are called angle strikes.

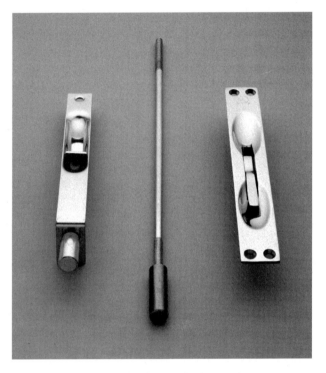

Standard flush bolts (left) are 6 in. long, whereas extension flush bolts (right) can be installed almost any distance from the top and bottom edge of a door. The only limitation is the length of the threaded rod (center).

Dutch-door bolts are surface bolts, too, but they have a special use. A Dutch-door bolt has to be short so that it won't interfere with the lockset mounted below or the dead bolt installed above the bolt. The strike on a Dutch-door bolt is angled (see the two strikes on the right in the photo above): The lip of the strike mortises into the top of the lower door, and the solid brass strike hole projects out beyond the face of the door, providing a strong, secure latching surface for the bolt.

Flush Bolts

Flush bolts are used on pairs of doors, where the inactive door must be locked at the top of the jamb and at the bottom of the sill or threshold so that the active door has a stationary surface in which to latch. Surface bolts can be used on stationary doors, too, but flush bolts are more common because they're hidden in the edge of the door's lock stile. Flush bolts must be mortised in flush to the front edge of the inactive door or the active door won't be able to swing past without rubbing.

Two types of flush bolts are used in residential construction: standard flush bolts and extension flush bolts (see the photo at left). Standard flush bolts are 6 in. long and are installed at the very top and bottom of the door. Extension flush bolts can be installed almost any distance from the top and bottom of the door, which allows them to be placed within easy reach, even on very tall doors.

Quality should be uppermost in mind when choosing flush bolts, because they're often exposed to moisture and some tend to rust. Besides, flush bolts take a lot of abuse, especially the bottom bolt on a pair of doors. Each time the active door closes the flush bolts get bumped; every time the active door is slammed the bolts are slammed even harder. But recognizing inexpensive imitations isn't always easy. My best advice is to stay with brand-name hardware like Quality, Ives, Baldwin, and Stanley.

Door Stops and Hold-Opens

Just as there is hardware that holds doors closed, there is hardware that holds doors open and hardware that stops doors from slamming into walls. There are an endless number of door stops and hold-opens (also known as door holders) available on the market. The most common stops are shown in the top photo at right. Spring stops are ubiquitous in many tract houses and have the advantage of surviving repeated attacks by vacuum cleaners. Rigid stops have a richer appearance and are preferred by some homeowners, though rigid stops are not forgiving to vacuum cleaners or to errant toes. However, rigid stops are the only choice if the door stop and the door meet at an angle. For situations where two doors swing back against each other, roller stops are excellent choices. They allow doors to open against each other without leaving bumper marks or dents.

Occasionally doors swing into cabinets, mirrors, or bathtubs, so hinge stops might seem to be the only choice. They're inexpensive, but they have expensive results: They force too much pressure on the hinge screws, the jamb, the casing, and the door. Eventually something has to give. A floor stop is a better alternative. Floor stops are available in many designs, both low domes for hard-surface floors and high domes for carpeted floors.

There are other door stops that serve a dual purpose. Door stop/hold-opens stop a door from banging into a wall and also hold a door open, even in a stiff breeze (see the bottom photo at right). I recommend these products in all windy locations, or for exterior doors that often remain open. Though they're more expensive than standard door stops, hold-opens save a lot of money in slamming-door damages.

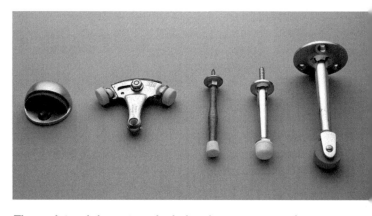

The variety of door stops includes the common spring stop (center), rigid stops in several sizes (second from the right), rigid roller stops for doors that swing back into other doors (far right), and dome stops that mount to the floor (far left). Only use a hinge stop (second from the left) if it's absolutely necessary because it often damages the wall, door, or hinge.

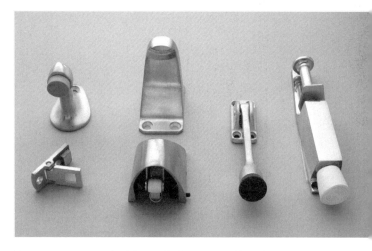

Door stops can double as hold-opens, like the heavy-duty commercial type (second from the left) or the lighter and more common residential series (far left). Old-fashioned drop-down hold-opens are one choice (second from the right), but I prefer the sleek Omnia retractable hold-open (far right), which can be extended or retracted without bending over.

CHOOSING HARDWARE 29

DEAD BOLTS AND LOCKSETS

Hold-opens and door stops are important, but dead bolts and locks are critical for security. Most dead bolts are designed similarly, but the prices really vary. And for some homeowners, selecting a lockset can be as difficult as choosing a new car because of the countless manufacturers and variety of products and finishes. In this section I'll tell you what you really need to know about dead bolts, then I'll work my way into bored locksets and mortise locks.

Dead Bolts

At the top of the price range are Baldwin dead bolts and the higher end Schlage B400, B600, and B800 series. These dead bolts are of the same quality, operate similarly, and share the same keyway (see the photo below) so they can be keyed alike (to accept the same key). Baldwin dead bolts are available in all major finishes and in several similar styles, and each is available with a decorative escutcheon plate at an additional charge. An escutcheon plate adds a look of elegance to an otherwise plain round cylinder. Schlage dead bolts are marketed in all major finishes, too, and a variety of escutcheon plates are available at an additional charge.

Westlock, Kwikset, and Schlage B100 series dead bolts are near the middle of the price range. Like Schlage and Baldwin, Westlock and Kwikset also share the same keyway—a Westlock dead bolt can be keyed to a Kwikset key. But Westlock and Kwikset keys will not work in Schlage or Baldwin locks. Kwikset has made a strong effort to re-engineer its entire line of locks, and its new Titan dead bolts are strong, work smoothly without hanging up, and have proven durable even in harsh environments like apartment housing. As with Schlage dead bolts, Kwikset and Westlock dead bolts are manufacturered in only one basic style, though different finishes are available.

Bored Locksets

There are two distinct categories of locksets: bored locksets and mortise locksets. Bored locksets are quick and simple to install. Usually only two holes have to be drilled for a bored lockset—one 2⅛-in.-diameter hole through the face of the door (called the face bore), and one 1-in.-diameter hole into the edge of the door (called the edge bore or latch bore). Bored locksets fall into two categories: tubular locksets and cylindrical locksets. Most bored locksets can be installed in less than two hours. Mortise locksets are an entirely different animal (see the photo on p. 34) and require far more tools, time, and expertise.

Tubular locksets Until the early 20th century, most locks were the mortise type, but shortly after the turn of the century tubular latches were introduced (see the top drawing on the facing page). The first tubular latches were strictly passage locks—they had no locking function like the buttons we push today on a bathroom door. Early tubular latches had very small rosettes, and the locks only required a ⅝-in.-diameter face bore. A handle, sometimes brass, porcelain, or cut-glass, threaded onto each end of a rectangular shaft that passed through the center of the latch. Tubular-latch locksets have changed a great deal since the turn of the century (Baldwin Type II locks are still designed in the same manner, however). Today, most tubular-latch locks

Schlage and Baldwin dead bolts are similar in design and come with dust buckets (in the black plastic). The Westlock dead bolt at the upper left is a double cylinder lock and requires a key on both sides.

have larger rosettes, require a 2⅛-in. face bore, and include privacy buttons for bathroom doors as well as key cylinders for entrance doors.

Cylindrical locksets In the early 1920s, Walter Schlage invented a completely new lock design, called the cylindrical lock (see the bottom drawing below). Schlage's lock has a keyway in the center of the exterior knob and a push button in the center of the interior knob that locks the device—most of us have pushed a Schlage button at least once in our life.

Regardless of whether a lock is tubular or cylindrical, there are four basic types of locksets. Keyed locks for entry doors and privacy locks for bathrooms and bedrooms are the only locksets that truly lock. Passage locks are used on closet doors and general-use rooms, but they can't be locked, only latched. Passage locks are often used on exterior doors along with a dead bolt. Some homeowners see no need to have two keyed locks on one door, while for other people two locks aren't nearly enough. The fourth type of lock is called a dummy lock

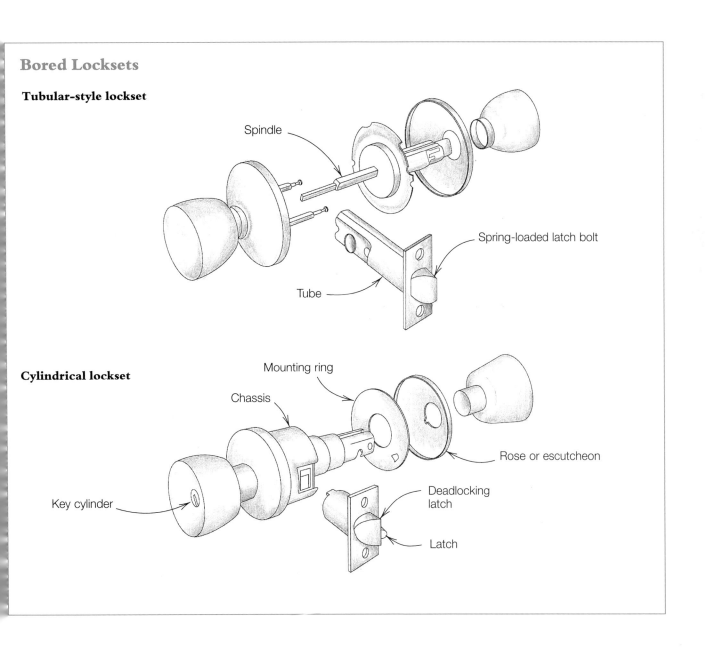

CHOOSING HARDWARE 31

Pocket doors require either privacy or passage locks. Choose a well-manufactured brand like the Ives 990 passage (left, on the bottom) and 991 privacy (left, on the top). Grodon manufacturers a pocket lock that fits into a standard lock bore (right). These locks work well and are easy to install.

because it can't even be latched—it's screwed to the face of a door in a fixed position and acts as a pull handle or a matching decorative handle on a pair of doors.

Schlage has several different lockset designs and price ranges. The A series lock is a heavy-duty residential design that utilizes Walter Schlage's original invention. The D series, for commercial applications and doors that have to meet handicap requirements, relies on the same design, though the D series is not available in many styles (see the photo on p. 25). Westlock makes a cylindrical lock that resembles Schlage's, though the locking push button on the Westlock nests in the rosette, or interior escutcheon plate, and not in the knob.

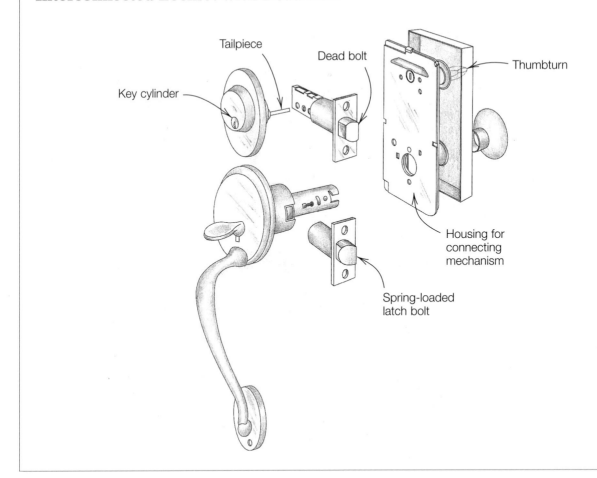

Interconnected Lockset with Dead Bolt

32   CHAPTER TWO

Both Schlage and Kwikset manufacture locks in the tubular design (see the photo on p. 25). The Schlage series F is a less expensive lock and is competitive in many ways with Kwikset locks. Baldwin manufactures a lower-priced (though not inexpensive) cylindrical/tubular lock line—the Images Series—in which the tailpiece is separate from both handles and slips into the latch from inside the door, before the handles are installed.

Pocket Door Locks

Pocket doors have an undeservedly bad reputation, primarily because many brands of pocket door hardware are constructed poorly. From the track to the wheels to the lock, pocket doors can be built to last if the right products are used (see Chapter 7). Traditional pocket door locks have suffered greatly because of cost-cutting and importation. The pot metal used for pulls and linkage inside some locks is brittle and begins to break down soon after the doors are installed. When purchasing pocket door locks, look for a reputable manufacturer. Ives offers both passage pocket pulls and privacy pocket latches in traditional styles, as shown on the left in the photo on the facing page. Traditional pocket hardware requires a rectangular mortise cut into the lock stile of the door, but a new lock is now available, manufactured by Grodon (shown on the right in the photo on the facing page), that uses the same 2⅛-in. face bore and 1-in. edge bore as a standard lock. This lock simplifies installation because it requires the same tools and procedure used for standard locks.

Interconnected Locks

Fire codes require panic-proof locking devices on exit doors: The dead bolt and latchset must be operated and opened with a single function, eliminating the standard combination of a separate dead bolt and latchset. Schlage and Westlock offer interconnected locks that operate in this fashion—turning the knob, lever, or key simultaneously retracts the latch and the dead bolt. The Schlage H series lock is a combination deadbolt and latchset that operates in unison through the addition of a special plate that mounts to the hardware on the inside of the door (see the drawing on the facing page). The Westlock 1500 series knobset and 1400 series handle set (shown in the photo at right) operate similarly through the use of an interconnecting trim plate on the inside of the door. Baldwin manufactures a panic-proof handle set in its Images line that also relies on an additional interior plate to interconnect the dead bolt and latchset.

As with all mechanical devices, more linkage often causes more problems. Interconnected locksets do not operate as smoothly as separate locks and dead bolts; often, unless the installer is particularly careful and the product free from all defects, more force is required to turn

Westlock 1500 series interconnected locks have linkage inside the interior trim to draw back both the latch and the deadbolt when the interior handle is operated. Likewise, when the key is turned, both the deadbolt and the latch retract.

CHOOSING HARDWARE 33

Lock Finishes

One of the most important recent advances in lock manufacturing has nothing to do with how locks are designed but rather with how they are finished. In the early 1990s Baldwin Hardware Corporation began experimenting with a new finish based on a high-tech process called physical vapor deposition (PVD).

In conventional metal plating, a thin layer of material, like brass or chrome, is deposited on the surface of a lock. If the lock is scratched or the lacquer finish degrades even slightly, tarnish can develop under the finish. Tarnish, like cancer, starts small, grows exponentially, and there's no cure once it penetrates a lacquer finish. PVD, as opposed to metal plating and lacquer finish, bombards another metal into the underlying brass of which the lock is made. The molecular bond that results from PVD is almost impervious: It can't be banged off, scraped off, or worn off.

Baldwin calls this new finish The Lifetime Finish. Like clad windows and doors, I've been recommending The Lifetime Finish for all exterior hardware, and particularly for locks near the ocean. Now Westlock, Weiser, and Kwikset offer the same finishes. Bear in mind that a PVD finish, like the lock itself, is only as good as the underlying substrate or brass. Baldwin locks are solid brass, and their price reflects that.

Mortise locks have been in use for centuries, like the old mortise lock at the top of this photo. Today mortise locks are more advanced and far more complicated (Baldwin mortise box in the foreground).

the key on an interconnected lockset than on a separate dead bolt. Bearing that in mind, I recommend Schlage H locks over Westlock interconnected locksets. Baldwin products are always dependable, particularly their Images locksets, though they never work as smoothly as Baldwin's mortise locks.

Mortise Locks

There's a big difference between bored locksets and mortise locks. Where bored locksets only require a simple 2⅛-in. face bore and a 1-in. edge bore, mortise locks necessitate a large and deep pocket mortise cut into the edge of a door, together with a special series of face bore holes that conform to the layout of the lock (see Chapter 6). Mortise locks (shown in the photo at left) are considered the Cadillacs of the hardware industry, and they rightfully should be. Mortise locks are the top of the line in quality and cost, and though some people consider them a nightmare to install, installed correctly they ride as smoothly as a luxury car.

The major manufacturers of mortise locks include Baldwin, Schlage, Jaddo, and Bouvet. The cases for most of these big locks look similar, but the locks are different.

Baldwin locks are the industry standard for residential use: Almost every manufacturer has imitated Baldwin in one way or another. Schlage mortise locks (known as L locks) are commonly found in commercial buildings. Look at any hospital, school, office, theater, or department store and you'll likely find Schlage L locks.

Baldwin mortise locks are available in so many trim designs that few stores carry a full display. Homeowners and decorators prefer Baldwin locks because an entry door is a major design statement and an entrance lockset is the focal point on many doors. But because these locks are available in so many configurations, Baldwin mortise locks require considerable forethought. Different size dead bolt cylinders, handle shafts, and even mounting screws have to be ordered for many types of trim, and the locks are not reversible. Each Baldwin mortise lock must be ordered for the correct hand of the door. Common among hardware manufacturers, Baldwin uses the "back to the hinges" system for handing their locks: Placing your back against the hinges, the direction the door swings is the hand of the door—if the door swings to the left it's a left-hand door, and if it swings to the right it's a right-hand door (see Chapter 6).

Because of their Spartan and utilitarian design, Schlage L locks are not commonly used in residential construction, which is unfortunate. Schlage L locks are workhorses. They go right in. They're easily reversed in the field (see Chapter 6). They never hang up. They never require call-backs for minor adjustments.

Jado and Bouvet, two European manufacturers, offer locks and trim designs that are substantially different from American designs. Jado's modern and sleek designs are easily identifiable. Bouvet produces entrance sets as well as lighter duty patio and interior mortise locks, which are similar to the old lock boxes found in turn-of-the-century homes.

Lock Strikes

Buying locks with matching keyways is an important detail whenever locks are replaced during a remodel, but matching lock strikes is almost more important. Lock strikes are mortised into the jamb leg opposite the latch, and strikes are manufactured in many different shapes.

If you purchase new locks, make sure the new strikes match the strike mortises in the old jambs. Schlage A and D series locksets, as well as Baldwin Type 11 locksets, are equipped with T-shaped strikes. Westlock, Kwikset, Schlage F, and Baldwin Images locks include a lip strike, also called a D strike. These strikes can be ordered with round corners (which facilitate installation in routed mortises) or square corners. A Westlock deadbolt strike is smaller than a Schlage or Baldwin dead bolt strike, though a Kwikset strike can be even larger. If the strikes that come with your new lock don't match your old strikes, don't give up hope. Some stores sell strikes separately.

Lock strikes come in different sizes and shapes. The T-strike (upper left) is used by Baldwin, Schlage, and Westlock in their higher-priced locksets. Lip strikes (both second from the left, one upside down) are available with square corners and round corners for ease of installation in routed mortises. Baldwin also manufactures one-piece latch and dead bolt strikes for locks that stack. Baldwin and Schlage dead bolts have even larger strikes (lower right), and Westlock dead bolts have a smaller strike (upper right). The thimble strike (lower left) is handy when adding a dead bolt to a door in a metal jamb.

CHOOSING HARDWARE

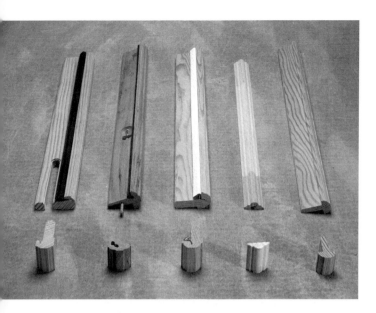

The common lumberyard astragal (far right) and the flat astragal (second from the right) come without weatherstripping. Astragals kerfed for foam weatherstripping are popular and effective (center). Astragals can be purchased with factory mortises for flush bolts, along with kerf-in foam (far left), though the flush bolts require additional fitting in the field. The Simple-T astragal has factory-installed extension flush bolts that don't require special fitting and Q-lon weatherstripping (second from the left).

ASTRAGALS FOR PAIRS OF DOORS

You need more than just a lock for a pair of doors. An astragal is necessary, too, so that the doors will not only lock securely against intrusive people, but seal against instrusive weather. Astragals for exterior doors are T-shaped. The top of the T acts like the door stop on a jamb and stops the active door as it swings shut. Nails or screws through the leg of the T fasten the astragal to the inactive door. Inactive doors are rarely opened and have to be latched with flush bolts or slide bolts in order to provide a stationary surface to which the active door can latch and lock—like the jamb leg to which a single door latches. Astragals are manufactured from many materials and in many different designs.

Lumberyard Astragals

I install wooden astragals more than I do aluminum ones. Most of my clients elect to stain their entry doors and choose a wooden astragal made from the same material as the door. A stained wooden astragal is a decorative addition to a pair of doors. Even if clients choose to paint their doors, they often want the astragal to blend in with the doors rather than making a separate statement in aluminum, bronze, or brass.

Wooden astragals are available from lumberyards, but lumberyards don't always carry the best assortment. The standard T-astragal found in most lumberyards has a narrow $3/8$-in. leg that's also beveled (see the photo at left). The reason for the bevel has eluded me and confuses most installers, too (see Chapter 6).

Lumberyards also commonly stock flat astragals (second from the right in the photo at left), though they're not normally used on exterior doors. Because they can only be glued and nailed to the face of the stationary door, flat astragals aren't as strong as T astragals, but for a pair of interior doors a flat astragal works fine.

Astragals with Weatherstripping

Like wooden jambs, wooden astragals are also available with kerf-in weatherstripping (center in the photo at left). All kerf-in foam products have admirable weather-sealing characteristics, and the additional space filled by the cushion of kerf-in foam allows more freedom for the seasonal movement of a door. Kerf-in thermoset foam can seal a gap up to $5/16$ in., so it's a perfect fix for doors that are slightly warped, twisted, or cross-legged (see Chapter 3). Wooden astragals using standard kerf-in foam should have a rabbet or flat area for the door that's $3/8$ in. wider than the thickness of the door. A narrow foam is also manufactured that only requires a $1^{15}/_{16}$-in. rabbet, but the narrow foam won't seal the same gap sizes as the standard foam.

Wooden astragals can be weatherstripped with other products, too. Silicone bead is one weather seal that's often used on astragals, but because there is such a wide range of seasonal movement between two doors and because pairs of doors are difficult to seal, I rarely recommend using silicone bead in an astragal. Silicone bead doesn't have the density or the gap-filling capacity of kerf-in foam. If a kerf-in product cannot be used, then cushion bronze is a better choice than silicone bead (for more on this, see Chapter 8).

Astragals with Flush Bolts

Flush bolts are often troublesome for installers, so to make life easier on carpenters some astragals are available with factory-installed flush bolts. Some weatherstripping manufacturers offer these astragals in oak, pine, fir, and primed paint, with factory-mortised holes for flush bolts (left in the photo on the facing page). These astragals are time-savers, even though the mortise usually has to be raised to accommodate the floor height, header height, or sill height. Occasionally inexpensive flush bolts are included with their astragals, but try to avoid using these. Choose an astragal that's supplied with substantial flush bolts. Pemko is one company that doesn't scrimp on this hardware.

One wooden astragal that comes with a true factory-installed flush bolt that requires no additional cutting or fitting in the field is the Simple-T (second from the left in the photo on the facing page). IDMM System manufactures this astragal with Q-lon weather seal and solid brass bolts that don't have to be moved to accommodate custom-size openings. These astragals are available in a variety of woods and lengths. Though the flush bolts are only $5/16$ in. in diameter (and require a very small hole in the head and threshold of an opening), I've never had a complaint about their strength or security.

Simple-T astragals have several advantages. The flush bolts are 10 in. long so that every astragal is equipped with extension flush bolts, which can be operated without bending down to the ground and without the need of a ladder or step stool for 8-ft. doors (on the top in the photo above). Also, Simple-T astragals are easy to fit and install—as you cut off the astragal with any carbide-tipped sawblade, the solid brass flush bolt is simultaneously cut. This product is available in all popular woods, as well as in paint-grade MDF.

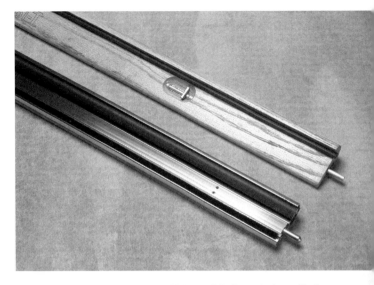

Aluminum astragals are available with factory-installed flush bolts, like Pemko's 3496 T-Bolt Astragal (bottom). These astragals are as easy to install as the Simple-T (top) and include two types of weatherstripping—a vinyl/foam seal and a pile inner seal.

Aluminum Astragals

Many styles of aluminum astragals are available from weatherstripping manufacturers. Most aluminum astragals have a narrow profile and depend on a slender vinyl fin to seal the gap between a pair of doors. These low-profile astragals are similar to rigid-jamb weatherstripping and do not include flush bolts. However, some aluminum astragals, like Pemko's 3496 T-Bolt Astragal (on the bottom in the photo above), include top and bottom extension flush bolts. This astragal is installed directly on the surface of the stationary door, while other aluminum astragals with integral flush bolts require a full-length dado on the edge of the stationary door.

Astragals like the Pemko 3496 are especially useful in harsh climates because of dual weatherstripping—both a vinyl-covered foam seal and a pile inner seal protect against air infiltration. In addition, the 3496 is manufactured with a thermal break to enhance the insulating quality of the astragal. The thermal break stops cold and moisture from traveling directly through the metal. These astragals are easy to install if the directions are followed closely, and few tools are required.

CHOOSING HARDWARE

Chapter 3
DOOR JAMBS

INTERIOR JAMBS

EXTERIOR JAMBS

When I was young, my family had a close friend who made his living as a magician. I was lucky because that man often entertained my childhood friends and me—for birthday parties and sometimes just for fun. I was lucky, too, because sometimes he showed us the secrets of his magic. The techniques always surprised me more than the tricks.

Carpentry is like magic because it depends on simple techniques performed in a precise order. This book follows that order, describing the step-by-step process of hanging doors and installing hardware. Some of these steps, especially in the following three chapters on jambs, doors, and prehungs, will be basic. Experienced carpenters are advised to skip around, but skip carefully! Along the way I'll include the tips and techniques I've learned that can give the craft an illusion of magic.

Doors can't be hung on air; they must have jambs. A variety of different jambs are used in residential construction. For the sake of simplicity I've separated jambs into two basic types: interior jambs for doors that are installed on the inside of a home, and exterior jambs for doors that lead out of a home.

INTERIOR JAMBS
With just two legs and a head—and only stud size and drywall thickness to worry about—determining the size of an interior jamb is easy. I use an interior jamb takeoff form to keep things organized and to ensure that I don't miss anything. When I walk a house or a set of plans, I always note the size of the doors first. Generally, rough-

Interior Jambs and Wall Conditions

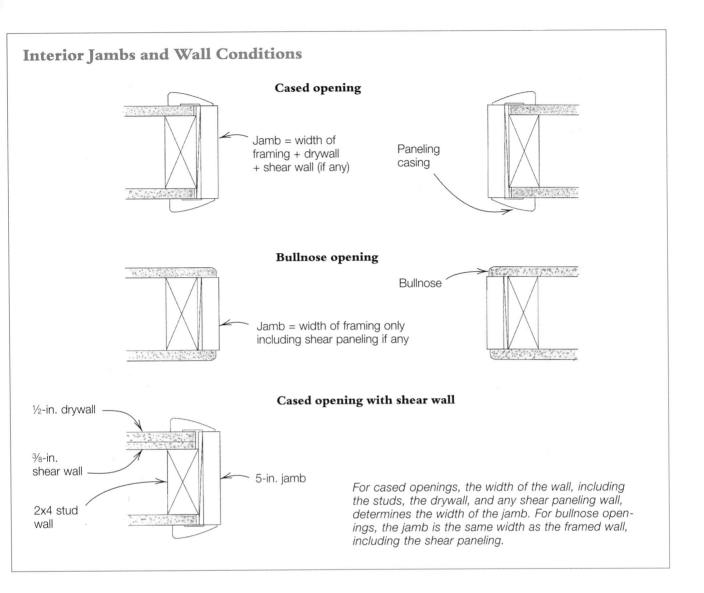

framed interior openings are 1½ in. to 2 in. wider than standard door sizes—an opening framed 2 ft. 10 in. is meant for a 2-ft. 8-in. door. Similarly, rough openings are framed 2 in. higher than standard door sizes—so a header framed at 6 ft. 10 in. is meant for a 6-ft. 8-in. door. The extra 2 in. allows sufficient room for the thickness of the jamb and some shimming. Because wall conditions can vary, I next check each individual opening on the job site or on the plans.

Most interior walls are framed with 2x4 lumber, so the thickness of interior walls is usually 3½ in., plus the thickness of the drywall. Although ½-in. drywall is most common, ⅝-in. drywall is also used in many homes.

This means that standard jamb stock can be purchased in 4 9/16-in. widths for use with 3½-in. studs with ½-in. drywall on both sides or in 4¾-in. widths for 3½-in. walls with ⅝-in. drywall on both sides.

Some interior walls are also shear walls, which means they're engineered to strengthen the structure of the home. The thickness of the shear paneling must be added to the width of the jamb to ensure that the jamb will be flush with both sides of the wall (see the drawing above). Shear panels vary in thickness, though often they are ⅜ in. or ½ in. A 2x4 wall with ⅜-in. shear paneling on one side and ½-in. drywall on both sides needs a jamb that's at least 4⅞ in. wide. Preferably the jamb should be 5 in. wide because shear paneling and drywall can never be nailed perfectly tight to the studs.

However, the millwork industry doesn't consider this factor when manufacturing standard-size jamb material. My supplier stocks jamb material in these limited widths: 4 9/16 in., 4 3/4 in., 5 1/4 in., 6 3/4 in., and 7 1/4 in. In theory these sizes are perfect. Because most homes in my area use 5/8-in. drywall, 5 1/4-in. jambs are popular for interior walls with 3/8-in. shear panels; 6 3/4-in. jambs are used on 2x6 walls with 5/8-in. drywall and no shear panels; and 7 1/4-in. jambs are used on 2x6 walls with 5/8-in. drywall and 3/8-in. shear panels. Whenever possible it helps to add an additional 1/8 in. to the width of all jambs to ensure that the jamb will be flush with the finished wall. This makes it easier to apply the casing later.

Assembling the Jamb

The head of a jamb is cut square at both ends and a factory rabbet is supplied at the top of every jamb leg so that the head can be fastened securely to the legs. The depth of the rabbets vary from 5/16 in. to 3/8 in., so the length of the head is almost 3/4 in. longer than the width of the door opening.

A work table the size of a full 4x8 sheet of plywood works nicely for assembling jambs, but an empty room with a flat floor works, too. Lay the legs out with the rabbets pointed in, and fit the head between the rabbets. Before nailing the jamb together, pull a tape measure across the inside of the head to ensure you're assembling the right size jamb. Then pull the tape measure down one jamb leg to make sure the finished frame will fit in the opening. It's easier to cut off the legs before assembling the frame, so do it now if the legs are too long.

I use a nailer to fasten jambs together. I hold the head tightly against the shoulder of the rabbet in the leg and shoot my first nail from the back of the leg into the end grain of the head, as shown in the photo below. Next I shoot a nail down through the top of the head into the top of the leg, which pulls the head down tightly against the shoulder of the rabbet. I continue in that order, alternating the nails and knitting the jamb together, driving three or four nails through the leg and two or three through the head.

Drywall screws pull two pieces of wood together tighter than nails, and screws are easier to use than hand-driven nails, especially since you only need two hands—one holding the jamb together and one running a cordless drill. But to drive nails by hand you need three hands—one holding the jamb together, one holding the nail, and one holding the hammer. Install screws in the same order as nails, but first drill pilot holes for each screw. Without a countersunk pilot hole, driving screws close to the end of a board will split the wood. I use a single-twist tapered countersink bit. You'll use fewer screws than nails because screws are stronger. Drive only two or three screws through the back of the jamb into the end grain of the head, and drive one or two screws through the back of the head into the top of the leg. That's plenty of fastening to secure a jamb so that it won't come apart while it's carried to the opening. But before the jamb can be installed, sometimes the trimmers must be set.

Lay the frame on a flat surface. Drive the first nail or screw through the back of the leg while holding the head jamb tight against the shoulder of the rabbet.

Setting Trimmers for Bullnose Walls

Typically, jambs for cased openings are installed after the drywall is hung because the casing is installed on top of the drywall, as shown in the drawing on p. 39. But jambs for bullnose walls are installed before drywall or plaster is applied because the corner beading ties into the jamb. When I install jambs for bullnose walls I set the trimmers the way I want them, so I ask the framers I work behind to tack the trimmers to the king studs and not nail them into the sills or the headers (tacked trimmers are easy to pry loose). To set the trimmers, first center them in the opening, then plumb each one perfectly and secure them with clinched nails.

Kerfing and Keying Jambs for Bullnose Walls

I cut a kerf or key into all jambs for bullnose walls so that the metal corner beading or plaster lath can be joined to the jamb. Kerfs and keys prevent future paint and plaster cracks between the jamb and the finished wall. Drywall kerfs ensure a straight and even reveal line along the jamb, and plaster keys provide a perfect termination point for hand-floated stucco and plaster.

Kerfs and keys should be ⅜ in. from the face of the jamb. A ⅜-in. jamb reveal allows enough room so that the hinge barrel isn't squeezed against the bullnose and the hinge pin can be removed without scarring the finished wall surface.

A table saw with a thin-kerf blade works for kerfing the edge of a jamb. Set the fence ⅜ in. from the blade and push the jamb through on edge, with the face of the jamb against the fence. Use a fingerboard to hold the jamb tightly against the fence, because jambs can be bowed a little. Bowed jambs drove me to distraction until I started cutting kerfs using a router fitted with a slot cutter. A router cuts a cleaner kerf than most table saws and will follow any contour.

I use a table saw with a dado blade to rabbet plaster keys, although a router chucked with a rabbeting bit will do the job, too. I cut plaster keys ⅜ in. deep because plaster lath is thick and requires a little more space to tuck in behind the jamb.

Kerfed jamb for bullnose wall

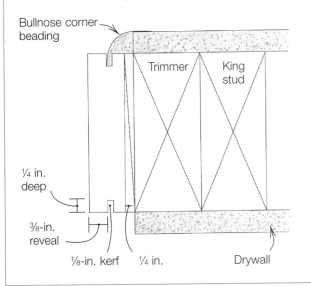

Keyed jamb for plaster bullnose

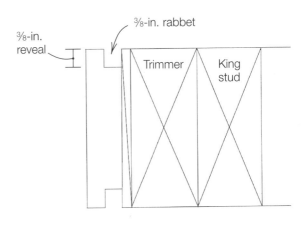

Centering a Jamb

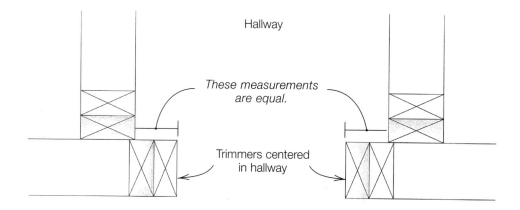

If the drywall hasn't been installed (above) and the trimmers can be reset, center the trimmers in the hallway opening. If the drywall has already been installed (below), center the jamb in the opening of the hallway.

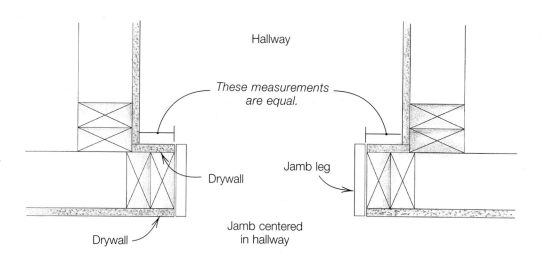

Center trimmers Because most openings are framed 2 in. wider than the door size and because jambs are generally only 1½ in. wider than the door size, there's about ½ in. of extra play that can be used to plumb trimmers. That extra ½ in. can also be used to center a jamb. Before prying loose either trimmer, look at the opening in relationship to adjacent openings and walls. If the doorway opens onto a hallway, the finished opening between the trimmers must be centered in the hallway, or else the margins between the door and hall walls will vary (see the drawing on the facing page). Likewise, if three doors are beside each other, the trimmers must be set so that the margins between the doors are equal. When a doorway leads into a hallway, mark the center of the hallway on the header above the door opening. Measure from that center mark back to the trimmers. Pry each trimmer loose and tack it to the header at the appropriate mark, but don't drive the nails fully home—adjustments might still be necessary.

If the doorway is one of a group of openings, find and mark the centers of the flanking openings first (see the drawing below), then measure between the centers of the two flanking doors to find the center of the middle

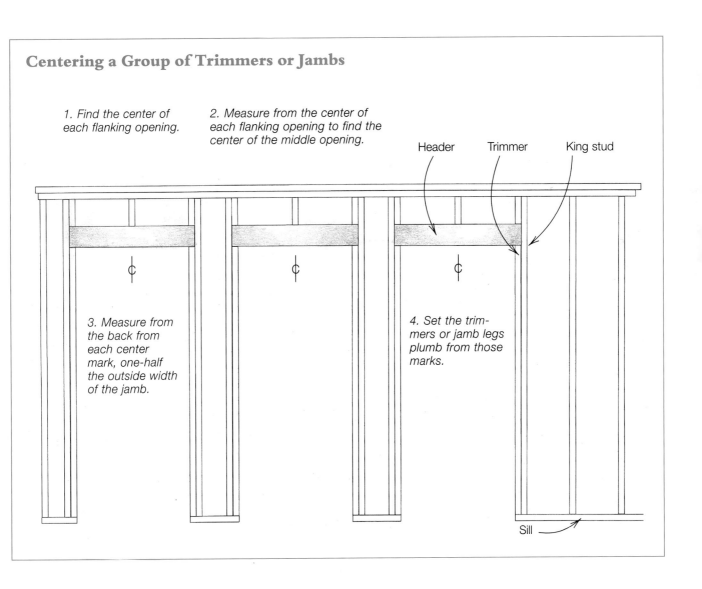

Centering a Group of Trimmers or Jambs

1. Find the center of each flanking opening.

2. Measure from the center of each flanking opening to find the center of the middle opening.

3. Measure from the back from each center mark, one-half the outside width of the jamb.

4. Set the trimmers or jamb legs plumb from those marks.

DOOR JAMBS 43

door. Measure back from those center marks half the width of each jamb. Tack the tops of the trimmers to the header at those marks, but don't drive more than a couple of nails in each trimmer—the openings might be way out of plumb and the trimmers may have to be shifted to one side.

Plumb trimmers I use a long level, tested frequently for accuracy, to plumb trimmers and jambs. A shorter level works well, too, if it's attached to a good straightedge, like a clean ripping of plywood or a perfectly straight 2x4 (a rare item). A slight bow or belly in the trimmer or jamb can be difficult to detect if the level or straightedge doesn't reach from the sill to the header or from the floor almost to the top of the jamb.

Clinched nails secure trimmers plumb and straight, and they save shims, too.

Once the tops of the trimmers are centered and tacked to the header, plumb and tack one trimmer to the sill. I use a tape measure to lay out the second trimmer because a tape is faster and more accurate than a level. I also want the trimmers set exactly to the outside width of the jamb. A tape measure ensures the jamb fits snugly in the opening.

Occasionally a rough opening is so far out of plumb that there isn't enough room to set the second trimmer. If there isn't enough room in the opening to plumb both trimmers, one of the king studs has to be knocked back and the sill cut off to create enough room. Use a reciprocating saw to cut the nails loose before banging on the trimmer and king stud.

Secure trimmers After measuring across the floor to locate the second trimmer, toenail it into the sill. Then nail off both trimmers to the header. Because the trimmers are now plumb you can ignore the bubbles and use the level as a simple straightedge. Occasionally I use shims to spread the trimmer from the king stud and bring the trimmer up flat against the straightedge, but clinched nails are quicker and they save shims (see the photo at left). Framers usually tack trimmers to king studs with 16d nails, so use a prybar to pull each trimmer flat against the level. Tack a 16d nail into the edge of the trimmer about halfway between the floor and the header, then bend the 16d nail over so that the nail head lies on top and near the center of the king stud. Drive an 8d nail into the king stud just above and inside the head of the 16d nail. Check that the trimmer is perfectly flat against the straightedge, then bend the 8d nail over the head of the 16d nail. Repeat the same nailing technique on the opposite side of the trimmer. More clinched nails may have to be added if the trimmer is still not flat against the level.

Installing Jambs

When my father built homes in Southern California in the 1950s, carpenters set trimmers exactly 1½ in. over the size of each door opening. The rough carpenters were also the finish carpenters, and they plumbed and straightened every trimmer, then set the jambs. That

Cross-Legged Openings

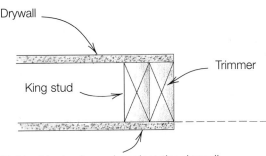
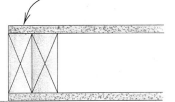

Drywall

King stud

Trimmer

Hold a block of wood against the drywall and hit this wall here with a sledgehammer.

Hold a block of wood against the drywall and hit this wall here with a sledgehammer to bring it in line.

These walls are not in line. The legs are probably not plumb. A door hung in this condition will not lie flat in the plane of the jamb, and either the top of the door or the bottom of the door will be proud of the jamb.

was well before the present age of framers and finish carpenters. In most parts of the country, framing crews now set trimmers 2 in. over the size of the opening, and they rarely use levels. For cased openings, like those in the top and bottom of the drawing on p. 39, interior jambs are installed after the drywall is finished, which means the trimmers can't be reset. For cased openings, there's no choice but to use shims—sometimes a lot of them.

Check for plumb Before placing the jamb in the opening, check that the trimmers are plumb. If the opening is slightly out of plumb, then the jamb can't be centered between the trimmers. Instead, the head of the jamb has to be closer to one side, and the bottom of the jamb has to be closer to the opposite side. To determine where initially to place the jamb in the opening, hold a level plumb against each trimmer and mark a plumb pencil line on the floor or on the header to indicate a plumb line. That mark also provides a way to check that there's enough room in the opening to plumb the jamb. It's easier to enlarge the opening before beginning work on the jamb.

Check, too, that the walls are plumb perpendicular to the trimmers. If the walls aren't plumb, especially if the trimmers are out of plumb to each other and out of line at the floor, then the opening is cross-legged, as shown in the drawing above. Use a sledgehammer or single jack to coax the walls over a little and bring them into line. Unless the walls are locked in place by lightweight concrete or concrete anchors, they'll probably move just enough. Hold a block of wood against the bottom of the wall to protect the drywall, and hit the block hard with the sledgehammer. Any damage to the drywall will be covered by the baseboard. If necessary, hit the other wall, too. Try to get the walls to align as best as possible, though minor adjustments can be made while setting the jamb. But I'll have more on that a little later.

DOOR JAMBS **45**

Shimming a Jamb

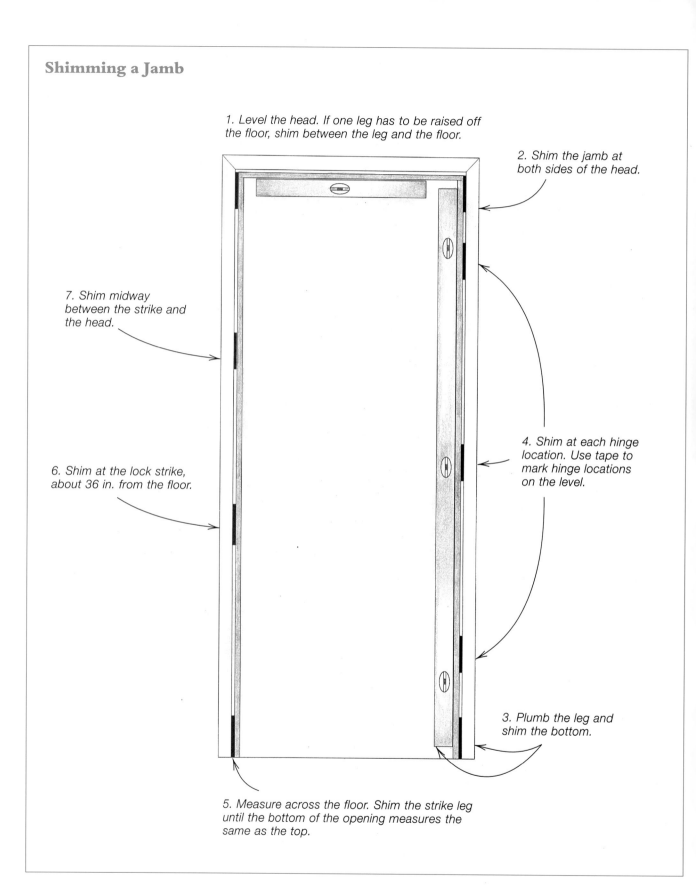

Level the head of the jamb Once you're sure the jamb can be plumbed in the opening, place a short level on the floor, from trimmer to trimmer. Usually the floor is out of level a bit and the level will have to be raised at one end or the other. Lift the low side of the level and slide a shim beneath it and against the trimmer. Now place the jamb in the opening. With one jamb leg sitting on the shim, the head of the jamb should be close to level. Hold the edges of the jamb flush to the drywall, with the jamb approximately plumb in the opening (close to the pencil line previously marked on the floor or header). Tack one nail near the head and one nail close to the bottom of each leg, in the center of the jamb. The purpose of the nails is to hold the jamb in the opening so that the head can be leveled perfectly, so don't drive too many nails. If you're driving nails by hand, back up the jamb with shims but don't nail through the shims yet.

Use a short level to check the head (as shown in the drawing on the facing page). Even though one leg was shimmed, the jamb may still require a little adjustment. Because the jamb is only tacked to the trimmers, a slap or two with the palm of a hand will level out the head. Make certain that the shim between the leg and the floor fits snug, especially if the hinge side of the jamb is raised off the floor. Without a shim between the bottom of the jamb and the floor, the weight of the door will work against the thin finish nails in the jamb, and the jamb will slowly settle to the floor. Eventually the casing will crack and the gap between the head of the door and the head of the jamb will grow. Fixing that problem is a lot more difficult than setting the jamb correctly in the first place.

Next, install shims at the top of each jamb leg, between the jamb and the trimmer. Keep those shims close to the top of the jamb, so they won't interfere with plumbing the legs. Use nails sparingly to just secure the shims, because the jamb may still have to be adjusted.

Plumb the legs To plumb and straighten the legs, start on the hinge side of the jamb and use a level that fits the opening or a good straightedge clamped to a short level. Hold the level or straightedge tight against the top of the jamb, with the bubbles perfectly centered. Pry the bottom of the jamb away from the trimmer until it

Use a long level to plumb the legs, and shim behind the hinges. Blue tape on my level marks the locations of the hinges on my router template.

touches the plumb level and make a pencil mark or drive a nail into the bottom of the jamb leg. One nail driven near the bottom of the jamb leg will hold the jamb still while it's shimmed square to the opening and then tight to the level. Drive one nail through the jamb and below the shims to hold everything in place.

I use blue tape on my long levels to mark the hinge locations on a jamb (see the photo above). I like to shim

DOOR JAMBS **47**

Cross-Sighting a Jamb

Stand here and sight past the opposite jamb leg. The jambs should be parallel and in the same plane, from the sill to the head. Keep one eye on the edge of this jamb here...

...and look up and then down the edge of this jamb here.

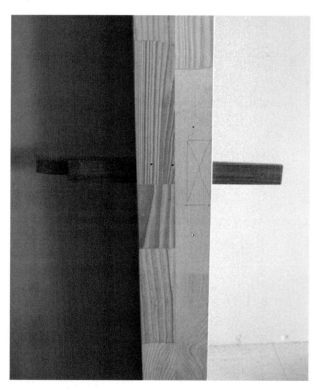

Keep nails away from the hinges, but be sure the jamb is well secured.

the jamb squarely behind each hinge, and bring the jamb flat against the level. But I don't like nailing too close to the hinge locations. The hinges have to be mortised into the jamb, and nothing ruins a sharp router bit quicker than a nail. Drive one nail at the center of the jamb below each group of shims. Repeat the same sequence at each hinge.

Rather than plumbing the strike leg, pull a tape measure across the floor, and mark a pencil line equal to the width of the jamb near the header. Shim the loose leg to that pencil mark, and tack a nail below the shims to hold everything in place. Install shims behind the strike location—about 36 in. from the floor—and midway between the strike and the jamb head. Shim the jamb leg until there's no daylight between the jamb and the straightedge.

Cross-sighting and cross-stringing Some call it cross-leg and some call it scissor, but either way it means the legs of the jamb aren't in the same plane. If the sledgehammer couldn't move the walls enough to bring them in line, then the jamb can be adjusted to take out al-

Cross-Stringing a Jamb

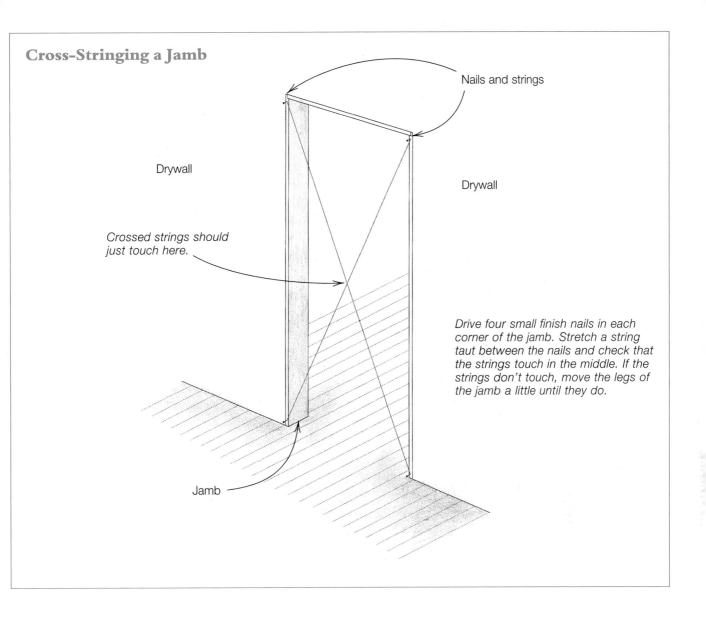

Crossed strings should just touch here.

Drive four small finish nails in each corner of the jamb. Stretch a string taut between the nails and check that the strings touch in the middle. If the strings don't touch, move the legs of the jamb a little until they do.

most ½ in. of cross-leg. If there's room beside the opening, stand back and sight from the edge of the jamb nearest you to the opposite edge of the farther jamb leg (see the drawing on the facing page). If there isn't enough room near the opening to allow a good cross-sight, then cross-string the jamb (see the drawing above). Drive a small finish nail in each corner of the jamb and pull a string from corner to corner. The strings should touch in the middle. If they don't, the jamb must be adjusted. A jamb should be adjusted for cross-leg at the sill, not at the head. It's a lot easier to fit the miters on the casing if the head of the jamb is flush with the finished wall surface. Use a block of wood and a hammer to tap each jamb leg a little out of flush with the wall, until the strings touch.

Once the jamb is plumb, straight, and cross-sighted, nail it off completely (see the photo on the facing page). Drive two nails through each shim, in a straight horizontal line, but stay back from the area of the hinge mortise. Drive another nail above and one below each hinge location.

Jambs with Transoms

Most jambs have two legs and one head, but transom jambs have two legs and three heads. One head, the head of the jamb, is at the very top of the jamb. The second head, at the bottom of the transom, is called the transom sill. The third head is the door head, at the top of the door.

Transom jambs can be built with continuous legs, in which case the sill of the transom is let into a dado cut in each jamb leg. Or transom jambs can be built from two complete frames that are the same width: an upper jamb attached to a standard door-size lower jamb.

Transoms that span wide openings for pairs of doors often sag. If there's room in the jamb design, I try to strengthen the transom sill by adding an additional piece of backing, like a small header, between the sill of the transom and the head of the door. Most often there's not enough height in the overall opening for additional backing. The best answer then is to embed a slender steel T-support between the transom sill and the door head.

Because a transom sill is normally covered with wood sash or direct glazed glass and glass stops, the top of the sill is rarely exposed. Ripping the sill down the middle allows room for two pieces of ¼-in. x ¾-in. x ¾-in. angled steel between the transom sill and door head, and rabbeting the bottom of each piece of the transom sill ¼ in. allows room for the thickness of the steel. A shallow dado, ⅛ in. to ¼ in., is needed in the top of the door head so that the steel and screws don't interfere with gluing the door head and the transom sill together. The two pieces of steel, sandwiched between the head and sill, form a rigid transom that won't sag and rub on the top of the doors.

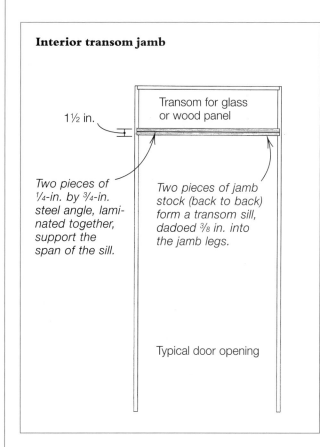

Interior transom jamb

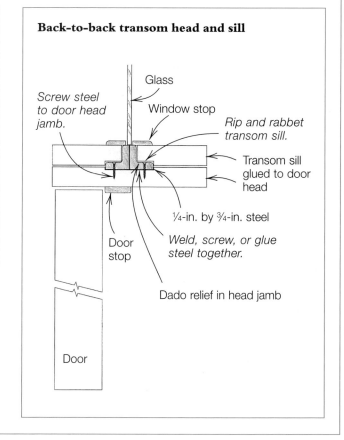

Back-to-back transom head and sill

50 CHAPTER THREE

EXTERIOR JAMBS

Exterior jambs are more complicated than interior jambs because they involve many more variables, like different size shear panels and insulation board, and a multitude of exterior wall finishes like stucco, wood siding, and brick. Exterior jambs often include a built-in sill or threshold, which makes a jamb for a swing-in door completely different from a jamb for a swing-out door. The addition of a sill or threshold also poses problems with header height and finished floor elevation.

To make working on exterior jambs more manageable, I've divided the topic into two categories: jambs with single openings for a single door or pair of doors, and jambs with multiple openings for single doors or pairs of doors with flanking sidelights and back-to-back mulls.

Sizing Single-Opening Jambs

Because measuring and laying out for exterior jambs involve so many variables, I keep a checklist of specifications about each opening. While discussing jambs with single openings, I'll review that list, but keep in mind that jambs with multiple openings share the same specifications.

The width Logically, the first thing you want to know is the size of the door or pair of doors. For a single door, the inside dimension (I.D.) of a jamb should be exactly the size of the door: The jamb for a 2-ft. 8-in. door will measure 2 ft. 8 in. from the inside of one jamb leg to the inside of the other jamb leg. The outside dimension (O.D.) of the frame will be 1½ in. wider than the I.D. because each side of the jamb at the door rabbet is ¾ in. thick (common jambs today are a little thinner than ¾ in. but it's easier to use ¾ in. when figuring the O.D. of a jamb). From this general rule some carpenters have concluded that rough openings for exterior doors are the same as those for interior jambs and should be framed 2 in. wider and 2 in. taller than the size of the door, but exterior openings are a lot different from interior openings.

If the opening is for a pair of exterior doors, always ask what type of astragal is planned (see Chapter 1 and Chapter 7). If a wooden astragal is being used, the width of the jamb and rough opening should be increased ¾ in. to accommodate the thickness of the astragal. Otherwise ⅜ in. will have to planed off the width of each door. Planing that much material off the edge of a door can be noticeable from more than one direction: When looking straight at the doors the lock stiles and hinge stiles will be different sizes. And planing ⅜ in. off the hinge stile will likely remove the veneer (most doors are veneered today and even the edges of the hinge and lock stiles have a ⅜-in.-thick solid wood veneer), which can be a major problem if the doors are to be stained.

The height Framing a door header 2 in. taller than the size of the door can also cause problems. Like the thickness of the astragal, the thickness of any built-in oak sill or aluminum threshold must be considered when determining the height of a jamb. As shown in the drawing on p. 52, the thickness of the oak sill (1¼ in.), plus the thickness of the head jamb (¾ in.), added to the door size (80 in.) equals 82 in. Therefore, an exterior jamb for a standard-size door will not fit in an opening framed 2 in. over standard-size door height.

I frequently encounter headers that are not framed high enough, and I'm often forced to build exterior jambs shorter so that they'll fit. Unfortunately, the I.D. of these shorter frames demands that the doors be cut down. I normally cut between ½ in. and 1½ in. off a door, depending on the I.D. of the jamb and the type of threshold and door shoe being installed. But cutting too much off the bottom weakens the door, which is why exterior headers should be set at least 2½ in. over door height, and even higher if the finished floor is a thick, hard surface, like some ceramic tile, stone, or hardwood floors.

It's best if an oak sill flushes out with a finished hard-surface floor (as shown in the drawing on p. 53) because the threshold will sit flat on both the sill and the flooring and span the joint between the two materials. But raising the sill (sometimes ¼ in. to ¾ in. so that it's flush with a hard-surface floor) requires even more header height, a fact few builders anticipate. The only alternative is to cut the jamb legs even shorter. But if the

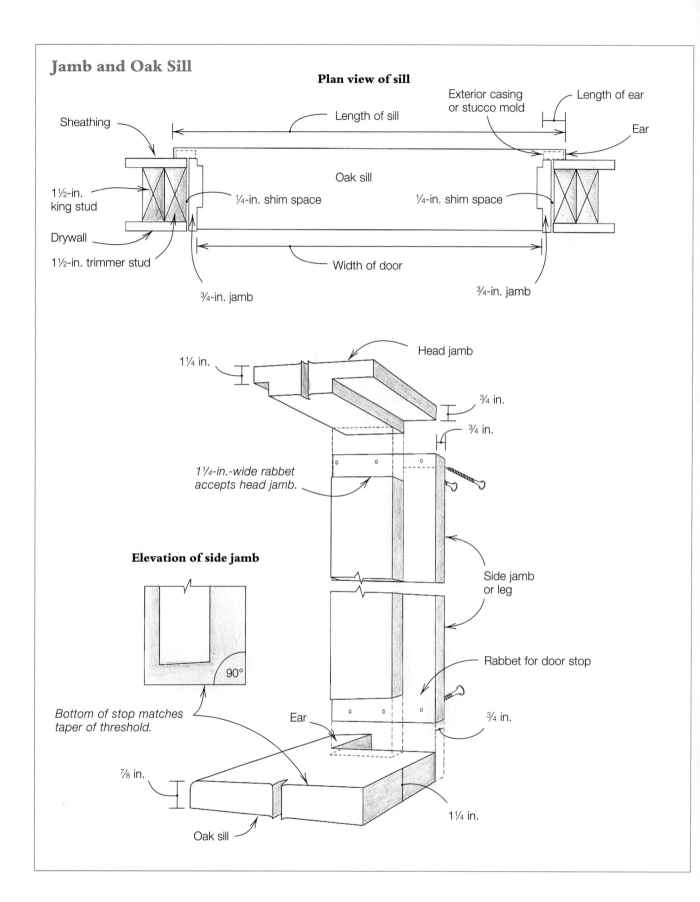

Raising an Oak Sill

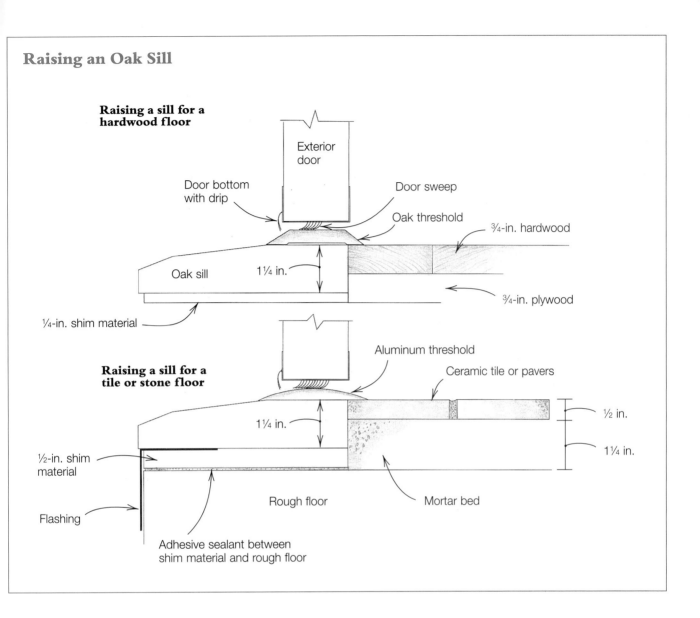

jamb legs are cut shorter, the door has to be cut shorter and the potential for weakening the door increases. For that reason, always measure the header height and ask what type of finished floor will be installed. On some jobs I've ordered custom doors for every opening because the exterior jambs couldn't be built tall enough and too much material would have had to be cut off the bottom of standard-size doors.

The wall condition As I said before, the wall condition for an exterior jamb is determined by many variables. The size of the trimmers is easy to see at a glance—trimmers are either 2x4 or 2x6. The shear wall isn't tough to figure either, if it's installed at the time you walk the house with a jamb checklist. When I walk a job for an exterior jamb order, I look at every opening and also check the plans. I look at the shear schedule and framing details on the plans to confirm the size and location of all shear walls because the width of the jamb has to include the shear paneling.

In my part of the country, most homes are finished with stucco, so exterior jambs are only built the width of the framed wall (including the shear paneling) plus the in-

Getting the Right Wall Condition

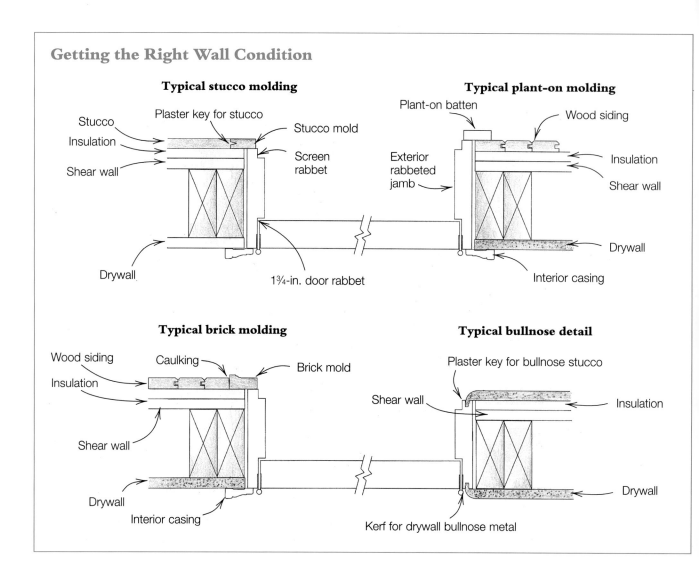

terior drywall or plaster. The thickness of the stucco is not included in the jamb width. Instead, stucco mold (exterior trim for stucco) is applied to the outside of the jamb, and the thickness of the stucco mold (⅞ in.) accounts for the thickness of the stucco (see the top left drawing above). The stock-size jambs I order are generally 4⅛ in. for 2x4 walls with ⅝-in. drywall inside or 4½ in. for 2x4 walls with ⅝-in. drywall inside and ⅜-in. shear outside. My supplier stocks exterior jambs in other sizes as well: 4¾ in. and 5¼ in. for 2x4 walls with additional shear or insulation board, and 6⅛ in. and 7¼ in. for 2x6 walls.

Jambs for wood siding require special widths, too, because there are many methods for trimming wood siding. One type of trim, called plant-on trim, is applied on top of the siding (see the top right drawing above), which requires that the jamb width include the thickness of the siding.

Another type of wood trim, brick mold or some other wood trim, is applied before the siding, with the siding abutting the trim (see the bottom left drawing above). A jamb for this type of siding and trim is ordered like the stucco mold jamb in the top left drawing—the width of the finished wall without including the siding

Exterior Jamb with Multiple Openings and Continuous Oak Sill

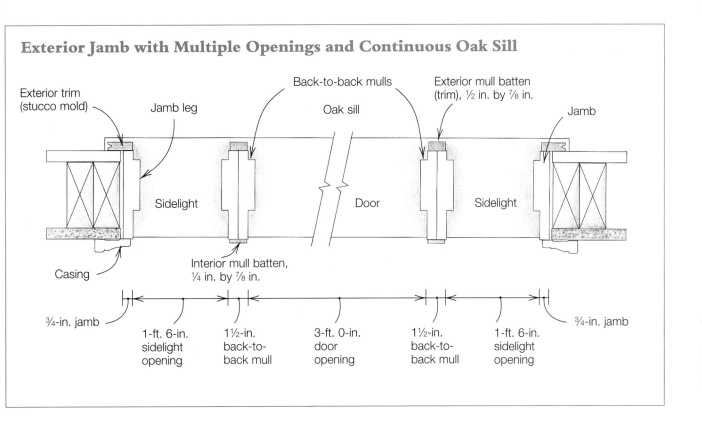

or trim. When buying a jamb for wood siding like that in the bottom left drawing, be certain the trim is thicker than the siding or the siding will be proud of the trim. If the trim isn't thicker than the siding, then order the jamb a little wider and shim behind the trim with a furring strip.

The jamb in the bottom right drawing on the facing page is a no-mold jamb, meant for bullnose exterior plaster and bullnose interior drywall or plaster. This type of frame is becoming more popular in the Southwest with a revival of the Santa Fe and Mission styles of architecture. Bullnose exterior jambs, like bullnose interior jambs, are the thickness of the final framing, including any shear wall or insulation board, but not including interior or exterior wall finishes.

Sizing Multiple-Opening Jambs

Though single-opening jambs and multiple-opening jambs share many of the same specifications (like wall condition, door size, and header height), the width of a multiple-opening jamb requires as much consideration as the height. In the accompanying drawings, I've used 1-ft. 6-in. sidelights with a 3-ft. door, but these openings can be a combination of many sizes, including pairs of doors. Regardless of the combination, jambs with multiple openings share one common feature: A pair of jamb legs always separates one opening from the next. Between each sidelight and the door is a pair of jamb legs. Each pair of jamb legs is also called a back-to-back mull.

When measuring and framing a rough opening, the width of back-to-back mulls is often confused and sometimes forgotten. I've encountered many openings framed for the combination of doors and sidelights, plus 2 in. for two jamb legs. This type of mistake causes hours of on-site jamb remodeling. Knowing the correct O.D. of a jamb is essential, and you can't determine that without knowing the inside dimensions. Each sidelight width should be the size of a standard door, like the 1-ft. 6-in. sidelights in the drawing above (1 ft., 1 ft. 2 in., and 1 ft. 4 in. are also standard sidelight sizes). The I.D.

DOOR JAMBS   55

of the center door or pair of doors should be a standard size, too, because custom doors are much more expensive than off-the-shelf doors. And remember to add the thickness of the astragal if a pair of doors with a wood astragal will be installed. For example, the I.D. of the center opening for a pair of 2-ft. 8-in. doors and a wood astragal should be 64¾ in.

Once the I.D. of the openings is known, determining the O.D. is simple arithmetic: Add the width of all the sidelights and doors (1 ft. 6 in. + 1 ft. 6 in. + 36 in. in my drawing), add the width of the two outside jamb legs (¾ in. + ¾ in.), then add the width of both back-to-back mulls (1½ in. + 1½ in.). Always write the O.D. measurement in inches only, so that no one confuses feet with inches. I used to mark openings for framers with measurements like 6 ft. 6 in. Sometimes I'd show up on the job with a jamb that measured 6 ft. 6 in. and a rough framed opening that measured 66 in.

The O.D. of a jamb can be even more confusing if the back-to-back mulls have been spread (see the drawing below). Often on remodels, I spread back-to-back mulls to fill slack space in openings that are too wide for standard-size doors. It only takes a little effort and imagination to spread back-to-back mulls and take up the odd inches. If an opening is 2 in. too big, I spread each back-to-back mull 1 in. apart so that the frame fits the opening perfectly. I also spread back-to-back mulls to add a more substantial appearance to a jamb and to allow for special trim applications, especially on openings of 8 ft. and taller. If the mulls have been spread, the spread has to be added to the O.D. of the jamb. A good formula for writing down the dimensions of a jamb that has back-to-back mulls looks like this (I am also including the wall condition so that once in the shop I have all the information I need):

1 ft. 6 in. by back-to-back mull (BBM)
with 1½-in. spread by 3 ft. by BBM with 1½-in. spread
by 1 ft. 6 in. by 6 ft. 8 in.

O.D. of jamb = 79½ in. wide by 81½ in. high
by 4⅛ in. wall condition

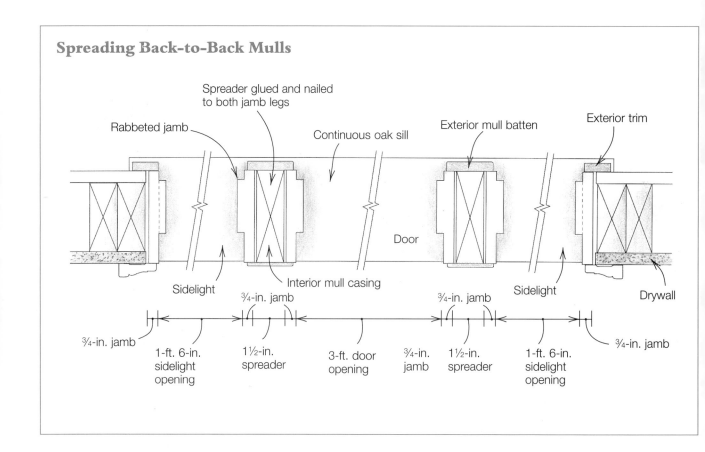

Spreading Back-to-Back Mulls

Full-Bound Sidelights

There are times when a continuous 8-ft.-long oak sill creates nothing but problems. Nearby landscaping and sprinklers can be one problem for a long oak sill, but too much exposure to the sun and the weather can also ruin the beauty of an oak sill, causing endless maintenance headaches. For that reason I also build jambs that don't have continuous sills beneath the sidelights. Instead, the sidelights stop between 3 in. and 8 in. short of the floor. A piece of rabbeted jamb—just like the head of the sidelight—completes the bottom of the sidelight and thus the name, full bound.

Properly made, full-bound sidelights are easy to trim. Raise the sill of a full-bound jamb high enough from the floor to allow room for a full size piece of baseboard beneath the sidelight sill. Install the casing as if the entire jamb were one door, with two full-length legs and one long head. Install mull battens on the back-to-back jambs, and finish the trim with a short piece of baseboard coped into the casing on one side and coped into the mull batten on the other side.

Exterior jamb with full-bound sidelights

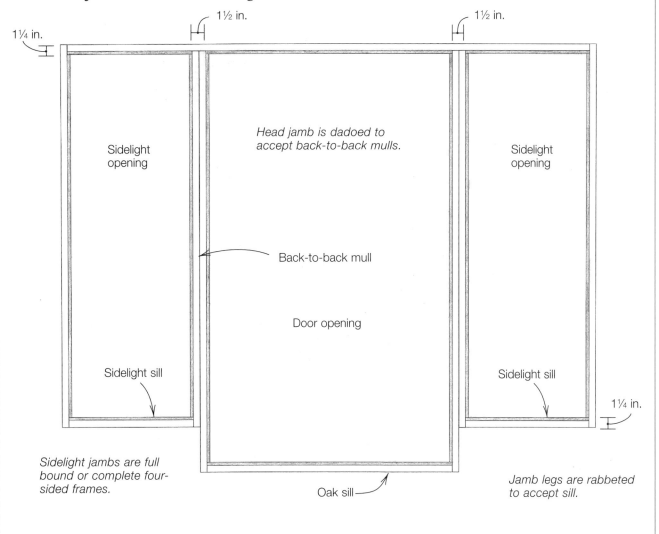

Door jambs 57

Lay out the ear the width of the jamb. The length of the ear should be slightly more than the width of the trim.

Use the sill to lay out the angled shoulder cut across a rabbeted jamb. Scribe a pencil line across the rabbet at the top of the sill.

Building a Single-Opening Jamb

As I've mentioned before, the process of carpentry is like the process of magic: Follow a simple, precise order and every aspect of door hanging works like magic, too. Building jambs is no different. Whether I'm building a jamb with a single opening or a jamb with multiple openings, I always begin with the sill.

Lay out the sill If the jamb includes an oak sill, I always measure and cut the sill first. The oak sill has to be as long as the width of the door opening, plus the length of both ears. The sill's ears should be a little wider than the width of the trim so that the trim can terminate cleanly at the sill. Once the sill is cut to length, measure in from one end of the ear the width of the trim and make a mark. Measure from that mark exactly the width of the door. The jamb legs will rabbet over the sill so these marks are the I.D. of each opening. The distance between shoulder cuts for each notch is exactly the width of the door, or doors if the opening is for a pair of doors. The depth of all the notches is the same, determined by the width of the jamb (see the top photo at left). I use a circular saw to cut the notch and a jig saw to finish the cut.

Rabbet the legs Like interior jambs, exterior jamb legs come with factory rabbets at the top of each leg, but because exterior jambs also have a factory-rabbeted door stop, one leg is the left side and one is the right side. And like an interior jamb, the exterior jamb head can be cut square and butted against the top rabbet in each leg. Jamb legs are also manufactured long, a leg for a 6 ft. 8 in. door is usually 82½ in. below the rabbet that's cut for the head. If the jamb doesn't require a sill, and an aluminum threshold will be installed later, then the jamb legs can be cut to the right length—short enough to fit beneath the doorway header—and the frame assembled. However, if the jamb includes an oak sill, the bottom of the legs have to be rabbeted to accept the sill (see the drawing on p. 52). When I build a quantity of jambs in the shop, I use a radial-arm saw fitted with a dado blade to cut jamb rabbets. But for building one or two jambs or for building a jamb in the field, a circular saw works just fine.

I use the sill to lay out the cut and to get the right angle across the shoulder of the rabbet (see the bottom photo on the facing page). Set the depth of the circular saw blade to the depth of the door rabbet (or screen door rabbet). Cut the shoulder first because it's the only cut that matters. Follow the line carefully so that the joint will be tight, then, with the circular saw set to the same depth, make additional cuts to take out the remainder of the stock. Finally, clean the rabbet with a wide chisel, as shown in the photo at right. I cut the jamb head last because it's the shortest piece of stock. Sometimes I can even cut the head from scrap material. The head is cut square, exactly the width of the door.

Assemble the frame Before assembling the frame, prime the end grain inside the notch of the sill. Moisture will probably find its way to the end grain, and because end grain is one of the most vulnerable spots on a sill, make sure that it's well sealed. (You can screw the legs to the sill while the primer is wet to create an even better seal.) Wait to prime the entire bottom of the oak sill until the jamb is completely assembled.

Make several passes with a circular saw to cut the shoulder, then carefully chisel out the rest of the material, using a wide, sharp chisel.

Drill a pilot hole through the back of the leg and into the edge of the oak sill. Use the first screw to draw the frame together before drilling pilot holes for more screws. Once the jamb is secured to the sill, drill a pilot hole and drive a screw through the bottom of the oak sill into the shoulder of the rabbet on the jamb to help stop the sill from cupping. One screw won't stop a wide oak sill from cupping, which is why it's important to seal the bottom of all oak sills. Seal the bottom of the sill and jamb legs before applying the exterior trim, especially if the trim is stain grade, so the primer doesn't get on the trim. Oak sills can be a problem because they're in contact with both the exterior weather and the interior heat and air-conditioning, so use at least two coats of exterior primer on the bottom of the sill.

Apply the trim Before the exterior trim can be applied, flashing—a waterproof membrane—has to be attached to the edge of the jamb. Flashing prohibits moisture from penetrating between the back of the jamb and the wall framing. A waterproof paper flashing is also woven into the water-resistant house-wrap system. Water-resistant house wrap collects any water that penetrates the siding or stucco and is supposed to shed water away from window and door openings. Eventually, water that penetrates the siding or stucco drains out at the base of the wall. Many new flashing products have begun to appear on the market. My current favorite is Moistop (made by Fortifiber Building Products), a kraft-paper product, similar to the old Sisalkraft paper I once used. Unlike Sisalkraft paper, which cupped, curled, and cracked before the siding or stucco could be applied, Moistop stays flat and holds up well. This improved flashing is made from kraft paper reinforced with glass fibers and covered on each side with a layer of weatherproof black polyethylene. Moistop comes in 6-in.-wide rolls that are 300 ft. long. The material is also available in 9-in.- and 12-in.-wide rolls, 300 ft. long.

Apply the flashing to the edge of the jamb legs first, using a small stapler. Hold the flashing back from the inside edge of the jamb $\frac{1}{8}$ in. more than the reveal

Scribe a line around the edge of the jamb, about ⅛ in. more than the reveal planned for the exterior casing. Staple the flashing at that line, first up the legs, then across the head. The flashing at the head should lap over both pieces of flashing on the legs.

Use galvanized finish nails to attach the exterior stucco mold or casing. Use glue and nails to hold the miters together.

planned for the exterior casing (see the top photo at left). If it's difficult to hold a straight margin by eye, then use a set of scribes to trace a pencil line on the jamb. Apply the flashing across the head so that it laps over the flashing on each leg.

Once the flashing is installed, the trim can be applied. If a jamb is being built in a shop, then trim the jamb there and save all the time it takes to set up tools on the job site. If the jamb is being built on the job, then all the tools must be out anyway. It's easier to apply the trim while the frame is lying flat in the shop. For doors that swing into a house, apply the trim flush with the screen rabbet. That way the width of the ⅜-in. screen rabbet is increased by the thickness of the trim and the jamb is prepared for future screen doors. For doors that swing out, hold the trim back from the edge of the jamb, leaving a ¼-in. reveal for hinge barrels. Miter the trim at the head, and on jambs with oak sills, bevel the bottom of the trim legs so they butt tightly against the sill. Attach the trim with galvanized finish nails, and use glue and nails to hold the miters together (see the bottom photo at left). After the trim is installed, fold any loose flashing onto the back of the jamb legs and use one or two staples to hold it in place during transportation.

Building a Multiple-Opening Jamb

As I said earlier, a jamb with more than one opening requires a back-to-back mull between each opening. I've explained how to cut and attach jamb legs to an oak sill for a single opening, but working with a back-to-back mull is a little more complicated. The sill requires special preparation for a back-to-back mull.

Lay out the sill Measure and lay out all oak sills the same way, whether they have single or multiple openings. Always begin at one end and measure and mark off the length of the ear. Then move your tape measure to the end of the ear, which is the beginning of the first sidelight (see the left photo on the facing page). If, as in my examples in the drawings on p. 55 and p. 56, the first sidelight is 1 ft. 6 in., make another mark at 1 ft. 6 in. Because each measurement is taken from the previous mark, move the tape again, this time to the end of the first sidelight, and lay out the first back-to-

Begin at one end of the sill and measure the length of the ear. From the end of the ear, measure down the sill the width of the first sidelight.

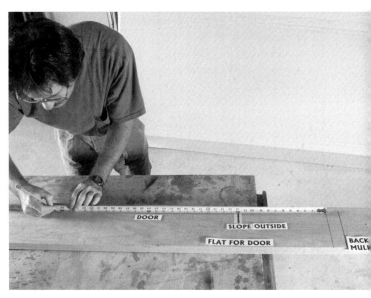

Move the tape to the end of the first mull and measure across the sill the width of the door opening, in this case 36 in.

back mull. The mulls in my example are 3 in. wide from door rabbet to door rabbet, so the sill has to be mortised to receive that exact thickness.

After laying out the mull, measure down the sill 3 ft. for the width of the door opening (see the top right photo). From that mark, measure over another 3 in. for the second mull, then lay out the last sidelight and ear. To check the measurements I always stretch my tape from the mark for the first sidelight jamb to the mark for the last sidelight jamb, then add 1½ in. for both jambs. In the examples in the drawings on p. 55 and p. 56, the O.D. of the whole frame should be 79½ in.

Cutting notches for the ears on a multiple-opening jamb is no different than cutting the ears on a single-opening jamb; but don't notch the sill for back-to-back mulls. All that notching in the middle of an oak sill weakens the sill too much. Instead, use a router and template to cut a shallow mortise in the top of the sill and pocket the back-to-back mulls into that mortise (see the bottom right photo). For more on mortising the sill, see p. 62.

A router template with a 1⅝-in. throat will cut a perfectly clean, flat mortise for a pair of back-to-back jamb legs.

DOOR JAMBS

Mortising Oak Sills

Cutting pocket mortises is quicker and easier than notching an oak sill; besides, a shallow pocket mortise won't weaken an oak sill as a notch will. The router templates I use to cut pocket mortises for back-to-back mulls are easy to make and can be used repeatedly.

Cut both of the legs the same height and slightly taller than the thickness of the oak sill. I make the legs on my templates 1⅜ in. tall because the sill stock I use is 1¼ in. thick. The length of the template is determined by the width of the oak sill, plus 1½ in. for the thickness of both template legs. The template I use for a 6¼-in. oak sill is 7¾ in. long because the legs on my tem-plates are made from ¾-in. material. The mortise is blind toward the outside of the sill and open on the inside of the sill, so the notch in the template is, too. The depth of the notch is only ¼ in., so any router can handle the job. The length of the notch is equal to the width of the jambs. In this example the notch is 4¾ in. long because the jambs are 4¾ in. wide. The width of the notch in this example is 1½ in. because the ¾-in. back-to-back mulls are laminated right to each other, with no spread between.

I've made several of these templates, each for different sizes of oak sill. But for mulls that are spread apart, all my templates are designed for a single-jamb leg. I slide the single-leg template from one mull location to the next,

Use a single-leg template to mortise for odd-size mulls. Cut the backing spreader ½ in. short of the jamb leg so the mull will seat completely in the mortise.

without mortising the area between the two legs. I cut the back-to-back spreader ½ in. short of the jamb legs so it won't interfere with the legs seating completely into the mortise.

Router template for back-to-back mulls

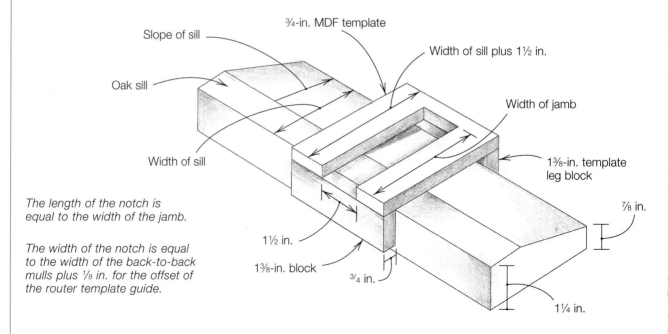

The length of the notch is equal to the width of the jamb.

The width of the notch is equal to the width of the back-to-back mulls plus ⅛ in. for the offset of the router template guide.

Dado the head I prefer to use a single continuous head for jambs that have multiple openings because it eliminates scissor joints between the jambs. The result is a stronger jamb. A continuous head has to be dadoed to accept each mull, but the layout and joinery are easy.

Because the head of the jamb butts into the rabbeted legs, the length of the head is exactly the I.D. of the total opening, just like the head of a single-opening jamb. If the oak sill is cut first, pull a tape measure from the shoulder of the notch for one ear to the shoulder of the notch for the opposite ear. That measurement is the exact length of the head. Laying out the dadoes in the head is the same as laying out the mull locations in the sill. Measure in from each end of the head the I.D. of the sidelights, so in the case of the examples in the drawings on p. 54 and p. 55, the amount would be 1 ft. 6 in. Mark a pencil line across the head at that measurement, which is the beginning of each mull. From that mark, measure across the thickness of the mull and draw another pencil line (the solid lines in the center of the photo at right). Repeat the same sequence on the opposite end of the head, then measure between the two mulls. The center opening measurement should match the sill layout and equal 3 ft.

To simplify the joinery between the mulls and the head, extend the mull layout lines ½ in. in each direction (see the dotted lines in the photo at right). This new layout will accommodate the total width of the mull (the first layout lines represented only the thickness of the rabbeted mull, not the complete mull). Cut the dadoes to the dotted lines.

Cut the mulls One step remains before the jamb can be assembled: The back-to-back mulls have to be cut to the correct length. It's easy to determine the length of the mulls by measuring one of the flanking jamb legs. Measure one jamb leg from the shoulder of the rabbet for the head, down to the shoulder of the rabbet for the sill. Add ½ in. for the depth of the mull pocket in the sill, and add another ½ in. for the depth of the dado in the head. That's the exact length of the mulls. Assemble the jamb in the same manner as a single-opening jamb, but drive four screws into each mull, through the head and through the sill.

If you cut the dado in the head to the solid dark line, the top of each mull will have to be rabbeted; but cut the dado to the dotted lines—about ½ in. wider in each direction—and the mulls won't have to be rabbeted.

Oak-Top Aluminum Sills

I occasionally use Tap-con concrete anchors to secure long sills to a concrete slab (for more see p. 67).

Adjust the depth-of-cut bolt on the sliding compound miter saw, then make repeated cuts until the notch for the mull is complete.

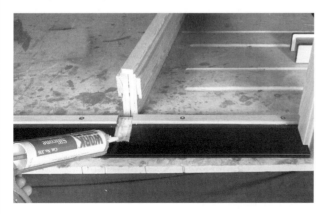

Apply a liberal amount of silicone adhesive sealant to the notch before attaching the jamb legs to the sill.

Because oak sills are milled from 6/4 lumber, they're expensive, and callbacks and warrantee work can drive the cost even higher, especially in climates where continuous oak sills are a nightmare no matter how well they're sealed. For that reason, the popularity of oak-top aluminum sills has grown (see Chapter 8). These sills are excellent alternatives for both single-opening and multiple-opening jambs with mulls. Many applications even eliminate the troublesome ears used on oak sills. Instead of the trim terminating on top of the sill, the sill terminates against the trim (see the top photo at left). The bottom of the jamb legs are rabbeted at a slope similar to the rabbet required for an oak sill; and the legs can be screwed securely into the sub-base of the aluminum sill, which usually consists of a wood substrate.

For jambs that have back-to-back mulls, the aluminum sill is mortised at each mull location exactly the same as the oak sill I described previously. However, rather than using a router to cut the mortises in an aluminum sill, I use my sliding compound miter saw. I adjust the depth-of-cut bolt on my saw to cut each dado deep enough that the mortises just open to daylight on the downhill side of the sloping sill (see the center photo at left). I repeatedly draw the blade back and forth until the entire cut is clean.

Cutting the bottoms of the jamb legs for an aluminum sill is no more difficult than cutting the jamb legs for an oak sill. I hold the jamb leg against the side of the sill and trace the slope of the sill across the rabbet of the jamb, just as I do for a solid oak sill (see the bottom photo on p. 58). To eliminate the possibility of water penetrating between the aluminum threshold and the substrate, be sure to apply a liberal amount of silicone beneath the mulls before assembling the frame (see the bottom photo at left).

To prepare the rough opening for the sill, lay a long level on the opening, and shim the level up so that the bubbles are perfectly centered.

Installing Exterior Jambs

The major difference between installing an interior jamb and installing an exterior jamb is the sill. If an exterior jamb has no sill, I install it almost exactly the same way I do an interior frame. However, if the jamb has a sill, the opening needs to be prepared before the jamb can be set. Preparing an opening always begins at the bottom. If the sill is installed out of level, then the frame will never be plumb and square. I can't stress this point enough. Almost every jamb problem I've had was the result of a sill installed out of level.

The sill Begin prepping the opening by cleaning the concrete or wood floor. Then use the longest level that fits between the trimmers. A short level placed on top of a straight board also works well. Shim up the low end until it's perfectly level, then fill in the gaps between the level and the floor with additional shims placed about every 12 in., as shown in the photo above. On concrete slabs, hold each shim in place with a blob of silicone adhesive. On wood floors, nail down the shims so that they won't move when sliding the frame in and out during the dry fit. Dry-fit all frames before applying adhesive or silicone sealant to the rough-opening subfloor, or slab. Silicone sealant is the best choice to stop water from penetrating beneath the sill, but I don't like it all over my hands and clothes.

Once the shims are positioned, slide the frame into the opening. If the header is too low or the trimmers too tight, then you'll have a problem, but that's not usually the case. Having answered all the questions on my order forms and premeasured the job carefully, I rarely have jambs that don't fit.

On-Site Jamb Remodeling

Mistakes happen when measuring and building jambs, and each time I have to cut a jamb down on the job I'm glad that it's fastened together with screws rather than with nails. Years of experience at cutting down exterior frames has taught me that it's easier to lower the head and recut the tops of the legs than it is to raise the sill and recut the bottoms of the legs.

To make the job easy, first remove the trim, then the flashing, and finally the head. Leave the legs fastened to the sill. After removing the head, cut the tops of all the jamb legs, then fasten the head back on, then the flashing, then the trim. Because the bottom of the trim is only beveled and not mitered like the top of the trim, it's easiest to cut the bevels again. Done.

The flashing pan Once I'm certain that the frame fits in the opening, I install a flashing pan or drain pan. Occasionally, general contractors supply sheet metal drain pans for exterior jambs, in which case I spread a heavy bed of silicone adhesive across the opening and press the drain pan down into the wet silicone. I use silicone to seal the ends and joints of the pan, and I drive any nails through the pan to secure it. Unfortunately, metal drain pans are more the exception than the rule, which is why I carry a roll of Moistop in my truck. I use several layers of Moistop to create a flashing pan that will seal the door opening. Cut the first layer of Moistop about 24 in. longer than the opening so that the ends extend up the trimmers. But before installing it, lay out a bed of silicone over the shims. Let a few inches of the flashing lap over the outside edge of the sill. Staples keep the flashing in place while the silicone adhesive is curing.

Cut a second layer the same length and bed it in silicone over the first layer, only this time let the excess paper flashing extend into the room. If the job is a remodel with hardwood flooring or a subfloor that the sill will butt against, wrap the excess flashing up the edge of the flooring to create a dam to isolate the sill from the flooring (see the photo below). If there is no flooring or subfloor, leave the excess flap until after the frame is installed, and then fold it to the inside edge of the sill to form the dam. Any excess flashing that projects up the sill and past the finished flooring can be trimmed later. Finally, apply a short piece of paper flashing into each corner, making sure that the stacked-up layers provide complete coverage. Before installing the frame, lay down another bead of silicone sealant on top of the paper flashing, but keep the sealant back from the exposed edge of the sill so that the silicone doesn't squeeze out when the jamb is pressed into place and create a real mess.

The legs With the unit in the opening and pressed down into the silicone, pull the sill tightly against the exterior wall and tack the jamb to the trimmers—down near the sill—to hold it in place. If the job is new construction with raw studs and no exterior siding or stucco in the way, straighten out the flashing before driving any fasteners through the upper jamb legs. On some remodels, the siding or stucco is normally removed to install a jamb. In such cases, the flashing has to be laced into the home's waterproof flashing system. On other jobs the jamb is flush with the finished siding or stucco wall, in which case the trim has to be set in a heavy bed of silicone to achieve a thorough seal (even then there had better be a sufficient overhang to protect the opening).

After tacking the frame in place, check the sill one more time to be sure it's level. Small adjustments can still be made while the silicone is wet. Once the sill is set correctly, pull the jamb tightly against the house and tack the top of each leg in place. There shouldn't be any gap between the back of the exterior trim and the face of the exterior wall. After tacking the legs, check the projection of the jamb inside the house all the way around the frame. I check this measurement carefully because before long I'll be back to install the casing and I don't want the drywall to be proud of the jamb.

The remainder of this installation follows the same procedure as that for an interior jamb. Use a long level that spans most of the leg from the sill to the head. Use a

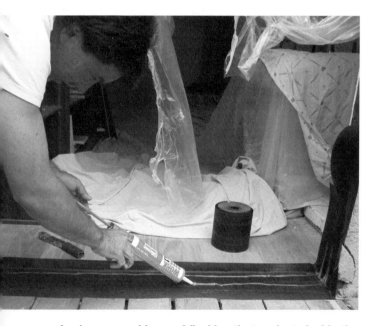

Apply a second layer of flashing that projects inside the house. Bend it up at the hardwood floor and create a dam to protect the finished flooring. Apply silicone on top of the last layer to seal the bottom of the jamb to the waterproof flashing.

short prybar to pry the jamb legs away from the trimmers and up against the plumb level. Once the legs are plumb, tack the tops and bottoms again, then shim behind the jamb until the door rabbet touches the level evenly along its length. Shoot a nail below each shim to stop the shims and the jamb from moving. I cross-string all wide openings because even a small degree of scissor can create a problem in a pair of doors (see the photo at right). It's still easy to move the jamb a little in the opening because it's not nailed off completely.

Once the jamb is adjusted for cross-leg and before driving any more nails, string the sill to be certain it's straight and level. Oak is stiff stuff and often refuses to lie perfectly flat. My favorite string for straightening sills and heads is 20-lb. braided Dacron fishing line, the strong, thin backing line that I use on my fly reels (available at any fishing store). I drive a finish nail in each jamb leg, in line with the top of the sill, then stretch my string tightly across the opening. I don't usually have to install more shims, but often I have to pull the sill down to the subfloor. On a raised wood floor a few nails will normally pull the sill down tightly to the shims.

On a concrete slab I drill holes in the sill for screws. First I drill a ⅜-in. counterbore about ⅜ in. deep, then I drill a 3/16-in. pilot hole through the sill. I use Tap-con concrete anchors to secure oak sills. These concrete screws are packaged with a special carbide masonry bit that drills the exact diameter hole required for the screw. After installing the screws I cover the screw heads with oak plugs. When I install aluminum sills with adjustable oak tops, I place Tapcon screws beneath the oak top. If I have to install a Tapcon in the face of an anodized aluminum sill or threshold, I spray the head of the screw with matching paint (see the top photo on p. 64).

Now that the jamb is adjusted for cross-leg, the sill secured straight and level, and the jamb legs tight against the exterior wall and perfectly plumb and straight, nail through all the shims near each hinge location. This is the same as with an interior jamb but with one small difference: Use galvanized or stainless finish nails to install exterior jambs. These nails won't rust or bleed through the final finish.

Cross-string all wide openings before nailing the jamb completely.

The head The head of a jamb has to be perfectly straight, and it's tough to get a long level inside a jamb. Jambs with multiple openings are especially difficult, so use the fishing line again. Drive small finish nails at the top of each jamb leg, in line with the bottom of the jamb head. Stretch a tight string between the jambs and shim the head right to the string. If the opening includes a pair of doors, place shims just off from the center of the head, so you won't have to drill a flush bolt hole through a nail. Getting a head straight is the last but most important step. When I install adjacent or adjoining exterior frames, I use a builder's level or transit to level and align the heads of all the jambs. On side-by-side openings it's crucial to keep the interior and exterior casing in line, just as it's imperative to maintain even margins between adjoining jambs. Sometimes there's more to installing a jamb than just plumb and level.

DOOR JAMBS **67**

Building an Interior Arched Jamb

Many aspects of carpentry appear complicated and inexplicable even from the short distance of front-row seats. But like magic, this trick depends on simple techniques. For instance, I build arched heads for interior jambs from three layers of ¼-in. MDF, but once the jambs are cased and painted you'd never know it.

If the wall condition is extremely wide (arched openings in mission style homes are often located in walls that are thicker than 1x12 lumber), I'll make the jamb legs from MDF, too, but for the legs I use ¾-in. material. Some carpenters cringe at the thought of an MDF jamb, but MDF jambs have been used for years in apartment and tract housing and have held up to the test. I conducted my own little test on MDF and found that, after drilling the proper size pilot hole, it takes almost the same amount of force to pull a screw from a piece of ¾-in. MDF as it does to pull the same size screw from a ¾-in. pine jamb. The fine fibers in MDF weaken quickly if a screw is driven in repeatedly, and for that reason I always use at least one screw in each hinge that's long enough to reach through the jamb and into the trimmer. But I follow that same practice for any jamb that carries a heavy door, whether the jamb is made from MDF or solid wood.

For arched jambs, use three layers of ¼-in. MDF so that the arch will be easy to bend. The walls of an arch are usually framed from two plywood templates, filled in with blocks ripped from 2x4s. These blocks provide excellent backing for the three layers of ¼-in. MDF to come. Rip the three layers to the right width, then cut them about 2 in. longer than necessary.

To measure the exact length of the arch, hold the first piece of MDF 1 in. into the opening, with the corner butted tightly against the trimmer. Don't try to put the entire piece of MDF all the way into the arch because it will be too long to fit. With just the corner butted against the trimmer, press hard against the MDF and walk your hands across the top of the arch to the other trimmer. If the opening is wider than a single door, borrow another pair of hands or tack a small finish nail through the stock to secure it temporarily. Mark the opposite edge of the arch. If the piece moved a little, or if you feel uncertain, cut it a little long. Fit it into the arch, trimming incrementally until it fits snugly against both trimmers and against the arched backing and blocking.

Before installing the first layer, use it as a template to cut the next two layers. Add ¾ in. to each succeeding layer of MDF. Later each piece can be trimmed to fit tightly in the opening, just like the first layer.

Spread glue on the backing blocks, then install the first layer. Fortunately, the first layer isn't exposed in the finished opening—the following two layers and the casing cover the first layer entirely. But get the first layer positioned tightly against the arch because it forms the base for the next two layers. Spread an ample amount of glue on the second layer and nail it to the first, then repeat the same procedure for the third layer. Nail the third layer sparingly because its face is exposed. Cut the tops of both jamb legs at a bevel to match the curve of the arch, then glue and nail the legs to the head. A little sanding is sometimes necessary to smooth out the three laminations and remove excess glue.

Building Exterior and Stain-Grade Arched Jambs

Exterior and stain-grade arched jambs aren't as simple to make as interior paint-grade arched jambs. Exterior jambs have rabbeted door stops, and stain-grade jambs can't be made from layers of plywood laminated against a header in the field. There are two methods used to build arched jambs, and both require a millwork shop. The first method, steam-bending continuous layers of wood around a form, is used by some manufacturers who turn out many jambs of the same size. This process requires ample shop space for steaming equipment and jamb construction, as well as form storage. The second and more popular method among small shops in my area is called bricking. In this process, short pieces of wood are cut to follow the radius of the arch, then glued together in a brick-like pattern, as shown in the photos on the facing page.

In the bricking process, the radius is laid out first, using adjustable trammels, on a large work surface and on two templates, one for the ¾-in. rabbet section of the jamb and one for the 1¼-in. door-stop section. A number of bricks are cut from each short length of 2x12 using a bandsaw.

Each brick is then screwed to the proper template and the rough cut of the bandsaw is finished by a shaper bit guided by a template bearing, which sizes all of the bricks identically.

The bricks are laid along the original radius and the ends cut to fit tightly then glued together in a brick-like pattern with the joints staggered evenly. Nails temporarily secure each brick in position, and great care is taken to place the inside surface of each brick smooth and flush with the preceding piece to minimize sanding.

The entire assembly is clamped and allowed to dry overnight before final sanding. Afterward, the legs are joined to the head in a sloping, rabbeted joint.

Chapter 4
PREFIT DOORS

PREHUNG DOORS

PREFIT BIFOLD DOORS

PREFIT BIPASS DOORS

MIRROR BIPASS DOORS

Ironically it's more difficult to set a jamb properly than it is to install a prefit door. Setting a jamb has to be done precisely using a tape measure, levels, and string to approximate the door and to ensure an accurate job. Installing a jamb with a prefit door, on the other hand, is easier because the door determines the position of the jamb. The process and techniques used to set and adjust a prefit door form the basic knowledge needed for hanging all doors, even those that have to be hinged, planed, beveled, and bored. For that reason this chapter on prefits precedes the following chapter on hanging doors the old fashioned way. Prefit doors are available for almost every opening. In this chapter, I'll begin with standard prehung doors, then discuss prefit bifold doors, and finally, prefit bipass and sliding mirror doors.

Though prefits can be built for almost every opening, there are instances when prefits are not the best answer—cases in which doors should be hung from scratch. Solid-core 8-ft. doors are one example because they're too awkward to transport and too unwieldy to set in a prehung jamb. Bullnose openings are another example, because the jambs often move after the drywall or plaster is applied, which demands on-site door hanging. But for all other interior doors and most exterior doors, prehungs are perfect.

PREHUNG DOORS
The real trick to prehung doors is ordering them correctly. There's a lot more to prehungs than just jambs and doors: wall condition and the swing (right hand or left hand); the type of hinge and the number of hinges;

the lockset, dead bolt, flush bolts, and more. And exterior prehungs have even more parts than interior prehungs. Many decisions that often wait until a home is nearly complete have to be made before ordering a prehung door. Let's look at interior prehungs first.

Ordering Interior Prehung Doors

Because there are so many aspects to a prehung door, I recommend using a list for ordering prehungs. I learned the hard way that lists are one way to make carpentry more enjoyable and more profitable. You can easily make your own prehung list. The first question, or column, should be the door size and thickness. The jamb type and width (see Chapter 3) should be noted beside the door size. Because prehungs include casing, the casing type and size have to be included with the order, too. One subject not usually of concern when setting jambs is the swing of the door, which is where most mistakes are made.

Handing Is it a right-hand door or a left-hand door? This question has puzzled builders, carpenters, and homeowners from coast to coast, and for a good reason: There are two conflicting methods for determining the hand of a door! Even in my area I have to use different methods to order doors from different companies. And on commercial jobs I often have to order a right-hand door from the door manufacturer and a left-hand jamb from the jamb manufacturer because each company uses a different method to determine the hand of the door. Maybe one day there will only be one method to determine the swing for all doors. Until then, it's a good idea for you to learn this:

The butt-in-the-butts method I teach this method to all newcomers, though some people consider it to be backwards. Put your backside (or butt) against the hinges (or butts) and use one arm to simulate the door swinging open. The arm you use is the hand of the door (see the top drawing on p. 72).

The hinge-side method Many people in the industry use the hinge-side method to determine the hand of a door. Face the door on the hinge side of the jamb so that the door swings toward you. If the hinges are on the left side it's a left-hand door; if the hinges are on the right side it's a right-hand door (see the bottom drawing on p. 72).

Installing prehung doors requires few tools, which saves time and money on the job site. But prehungs have to be ordered carefully to avoid costly mistakes. Solid-core prehungs, like the one shown here, typically come with the door separate from the jamb sections.

PREFIT DOORS 71

Determining the Hand of a Door

The butt-in-the-butts method

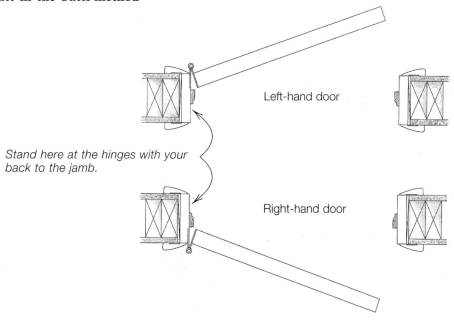

The hinge-side method

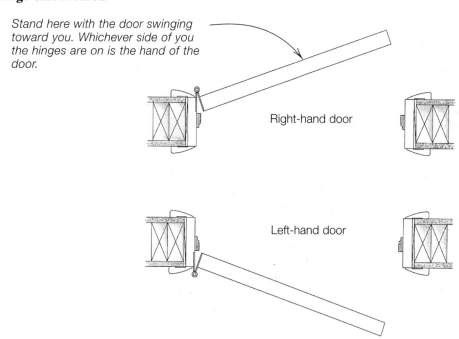

72 CHAPTER FOUR

Although these two methods are the most common, the reverse bevel is another designation for the hand of a door. Doors that swing outside a building are commonly called right-hand reverse or left-hand reverse doors. However, these terms mostly apply to locking hardware and not to prehung doors.

How do you know which method is right? You can't because there is no right method. So when ordering prehung doors, always specify the method used to determine the hand to make certain that you and the door supplier are using the same system. I send in my prehung orders by telephone facsimile, but I call the salesperson to confirm that, for my order, I'm standing with my butt in the butts.

Lock preparation Before a prehung can be ordered, the type of lock has to be selected and the backset has to be determined (for more on backset, see Chapter 2). Most locks require a 2⅛-in. hole bored through the face of the door, centered 2⅜ in. back from the edge of the door. This measurement is called the backset. But some locks use a smaller face bore and have a different backset. The jamb has to be prepared for the lockset strike, too, which varies in shape (see the photo on p. 35). If you make all your lock decisions before ordering a prehung, the locks can be installed quickly and easily after the door is painted.

Ordering Exterior Prehung Doors

A prehung list should include every door in the house that can be prehung, which includes just about everything, even exterior doors. Besides the jamb condition and jamb material, note the type of exterior trim the door-hanging shop should install on the exterior jambs and whether the trim is brick mold for wood siding or stucco mold (see Chapter 3). If the prehung manufacturer is applying the exterior trim, have the company install waterproof paper flashing on the jamb, too (see Chapter 3).

Exterior prehung doors should be ordered prebored for dead bolts, and the distance between the lock and dead bolt—called the spread (4 in. to 6 in. is standard)—should be specified. Always be sure to order NRP hinges for exterior doors that swing out. (See Chapter 2 for more on NRP hinges.)

Pairs and sidelights Doors for multiple-opening jambs can be prehung too. I don't like lugging all my tools up long flights of stairs, so I prefer prehungs for all doors in apartment buildings and commercial buildings and for any opening that's up several flights of stairs, whether for a single door or for a pair of doors with sidelights. In some cases, I order the prehung unit knocked down, so that I can screw the jamb together once I have

Exterior prehungs often include weatherstripping and exterior trim, though this one was ordered bald, with no exterior trim.

Knock-down prehungs are strange but simple. The four separate pieces include the door, with half the hinges attached, the hinge jamb, with the other half of the hinges, and the strike and head jambs. Each jamb section comes with casing installed on both sides.

it near the opening. In other cases, the jamb is delivered complete, with the trim and waterproof flashing installed.

Single-opening prehung units are straightforward, but even more components have to be specified with a multiple-opening prehung unit. As in the jamb take-off for multiple openings in Chapter 3, the space between the back-to-back mulls has to be defined (especially if a structural post has to fit within the mull pockets) and the outside dimensions of the unit have to be exact. The trim for the mulls and the locking device for active sidelights should also be included.

Weatherstripping Weatherstripping is included on many exterior prehung doors (see Chapter 8 for more on this). Specify all the weatherstripping, both the threshold type and color and the door bottom type and color. For prehung exterior doors, I like to install a sill nosing or sill cover beneath the threshold. I always ask the prehung manufacturer to send the nosing loose along with the doors so that it matches the color of the threshold.

Having a sill nosing is important because, like an exterior jamb, a weatherstripped exterior prehung door has to be shimmed high enough to clear the interior finish flooring (see the drawing on p. 53). Even an exterior prehung swinging over a vinyl floor has to be shimmed up high enough to clear future underlayment, vinyl, and a throw rug. When I see an exterior prehung unit with a built-in threshold and door shoe, I instantly want to know what the finished floor elevation will be. It's critical to measure the height of exterior openings carefully and know the finished floor covering so that all necessary shim material can be subtracted from the height of the rough opening before the prehung unit is ordered.

While on the subject of weatherstripping, most exterior jambs are now prepared for kerf-in weatherstripping, such as Q-lon, so when ordering your door, be sure to specify kerfed jambs. On pairs of doors, specify whether the astragal is wood and kerfed for weatherstripping or is aluminum. Many aluminum astragals are manufactured with integral flush bolts, but if the astragal is wood, remember to order it prepared for flush bolts; some are not and these flush bolts must be installed in the inactive door (see Chapter 2 and Chapter 6).

Custom doors Custom-size prehungs are also manufactured regularly. Often exterior mechanical rooms, water heater enclosures, and storage rooms have non-standard openings. So explain in writing on the order whether there are custom sizes or configurations. The notation "full bound" is a common term for a jamb that wraps completely around the door, with a sill identical to the jamb head. This is typical for a short furnace door or for a scuttle door to an attic space.

Installing Prehung Doors

There are two distinctly different methods for installing prehung doors. One method is for hollow-core doors, which don't weigh much, are easy to place and adjust in an opening, and don't require substantial shimming. A second method must be used for solid-core doors because the weight of the door tends to deflect the jamb, which causes adjustment problems that make it more difficult to set the unit. Exterior prehungs are set in much the same manner as solid-core doors, though additional attention is required because of integral thresholds. First I'll cover hollow-core prehungs, then solid-core doors, and finally exterior units.

Installing hollow-core prehungs Most hollow-core interior prehungs arrive on the job site with the casing attached to the hinge side and three loose pieces of casing that have to be nailed onto the stop side after the jamb is installed in the opening. Solid-core prehungs, on the other hand, normally come without casing attached (see the photo on p. 71). Other hollow-core units, called knockdown prehungs, come in four pieces (see the photo on the facing page), with the door separate from the three jamb sections. Each of the jamb sections has casing attached to both sides. Split-jamb prehungs are also available in some parts of the country. Split jambs look like they've been ripped right down the middle—one half of a split jamb includes the hinges and the door, with the casing applied; the other half of the jamb, also with casing, includes the door stop (see the photo above).

Split jambs adjust for different wall conditions, which is an outstanding feature. The two jamb sections meet at the center of the wall in a tongue-and-grove joint. The line of the joint is hidden by the door stop.

Layout and preparation Across the country and throughout the industry there are many different methods for installing hollow-core prehungs, and many work well. Some carpenters first check the trimmers for plumb and the walls for cross-leg (twisted or out of parallel trimmers), then check the sill for level and the header. But being raised in production-oriented southern California, I prefer to get right down to business.

Before installing any doors, I walk the house and mark an X on the wall beside each door opening. The X tells me which side the hinges are on. The casing will cover the X later. Next I take all the doors to their correct openings and check each opening to be sure it's framed 2 in. over the size of the door. I don't install any door until it is leaning near its properly sized opening and every opening has a door. It's easier to make last-minute changes and move the doors around before they're nailed in place.

Whether the unit is prehung and arrives on the job sight with the casing attached to one side or is knocked

PREFIT DOORS 75

Knock-down jamb heads have to be installed first. Hold the jamb head at a slight angle and slide it up and over the drywall. The casing wraps around the wall and will hold the jamb head securely in place while the legs are installed.

Tip the top of each jamb leg in toward the back of the head jamb. Simultaneously tuck the top of the jamb leg in behind the jamb head while sliding the casing over the wall.

As the jamb leg slips onto the wall and becomes vertical, the lock joint between the jamb leg and the jamb head will engage and tighten.

down with casing on both sides, the object is to get the unit in the opening—quickly. For prehungs that are cased on one side and for split-jamb prehungs, this is an easy task. Usually only one or two temporary nails or screws are needed to secure the door in the jamb. But before placing the unit in the opening, be sure to remove those temporary fasteners. Chick's Door Plugs are fast becoming the standard temporary fastener for prehung units. These devices fit through the lock bore and unthread from the back side of the jamb, which eliminates holes in the jamb caused by temporary nails.

Initial installation of knock-down prehungs takes a moment longer. Each of the three pieces has to be installed separately, beginning with the jamb head (see the photos on this page). Once the jamb is installed, the door can be swung on the hinges.

The nailing sequence Sometimes a prehung door is shipped to the job site bald, without any casing attached. This is usually true for openings that require wide or elaborate casing that's too delicate to be shipped attached to the jamb. Before attempting to fasten the unit in the opening, be certain the casing is securely attached to the hinge side of the door.

Once the unit is in the opening it might seem that you need to have four hands, but two hands and one foot work fine. First, look at the hinge side of the jamb. If there's no adjacent wall or door to use as a guide, hold a level against the casing on the hinge side and tip the frame until the hinge side is plumb. Use a prybar under the door to balance the door and jamb, then shoot a nail in the hinge side casing through the drywall and into the trimmer, about 3 in. below the miter. If you don't have a pneumatic nailer, place a shim under the bottom of the door to hold the door steady while you drive the nail by hand.

Step back and look at the head gap (the margin between the top of the door and the jamb head). The head gap should be about the thickness of a nickel. If the head gap is too tight, try and lift the strike leg a little. Don't nail the strike leg yet; just be sure there's enough room to raise it. If the head gap is too big, place the prybar beneath the door on the hinge side and raise the door straight up until the head gap is about the thickness of a nickle. The jamb will still move because only one nail has been tacked through the casing. Hold the jamb in that position with the prybar or shims, and drive another nail through the casing about 3 in. below the bottom hinge. Check the hinge jamb for plumb again. If it moved, use a hammer and a small block of wood to tap the jamb back to plumb. It only takes two nails to hold the door in position, which means the door can still be adjusted.

If a door is standing alone, without another door or wall nearby, then a good level should be the only tool you need to judge the door's position in the opening—perfectly plumb and level is always the best choice. However, if there are adjoining walls or doors, other elements have to be considered when positioning a door in a rough opening. Before driving any more nails through the casing or jamb, check that the margins between adjacent doors and walls are even, as shown in

Check the casing on the jamb in relationship to nearby casing and walls. Those margins should have equal measurements from top to bottom.

the photo above. The top of the door casing should line up with adjacent door and window casings. (Sometimes I shim a door off the floor so that the head casings will line up perfectly.) Also measure from the back of the casing to the wall or adjacent casing. Check near the top and bottom of the door. The jamb is still loose, especially the strike side (it has no nails in it yet), so hold the jamb with proper strike gap—about a nickel's thickness—while measuring the distance to adjacent walls or door casing.

Straightening a tweaked jamb The weight of a door pulling down on the top hinge can cause the jamb to deflect. Deflection at the top hinge, or tweaking, usually is a problem that occurs on heavier, solid-core doors, but it can cause minor problems with hollow-core doors, too. Deflection results in a wide gap between the door and jamb above the top hinge (see the drawing on p. 78). Such a large gap pushes the top of the door too close to the strike jamb, which in turn creates a large

PREFIT DOORS 77

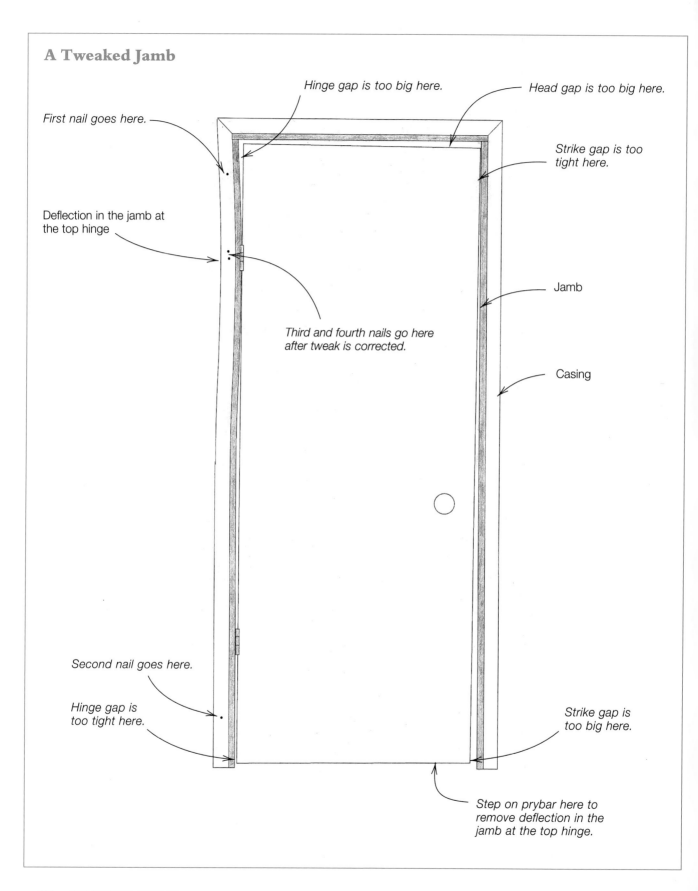

gap between the door and the jamb head above the lock stile. The nailing sequence is important because proper nailing order effectively straightens a tweaked jamb. After nailing the hinge jamb near the top and near the bottom, place a prybar under the leading edge of the door and step on it lightly. Apply slightly more than enough pressure to take the weight of the door off the jamb, so that the jamb, behind the top hinge, bends back a little toward the trimmer. While stepping on the prybar, I like to have a tight gap between the door and the jamb above the top hinge and a slightly larger gap between the door and the jamb beneath the bottom hinge. After driving two nails through the casing behind the top hinge, which is the next step in the nailing sequence, the door will settle slightly and the gaps will be perfect.

Once the hinge side of the jamb is set perfectly plumb (or straight to adjacent openings), make final adjustments in the head gap. If necessary, lift the strike leg a little to get a gap the thickness of a nickel, then drive a nail through the casing about 4 in. below the miter. Fasten the strike jamb with a nail every 16 in. or so, always checking the margin between the door and the jamb. Be sure to drive one nail through the casing and into the trimmer near the lockset. Also drive a nail through the casing and into the trimmer at the bottom of each jamb leg, so that if someone cuts a piece of baseboard a bit too long it won't push the jamb and ruin the consistent gaps.

Adjusting for cross-leg Open and close the door to be sure it clears the jamb and isn't rubbing anywhere, then step through the opening and shut the door against the door stop. Sometimes the door won't lie flat against the door stop on the strike jamb, which means the opening is cross-legged and the legs of the jamb need to be adjusted. If the door is touching the stop on the strike side at the top of the jamb but not at the bottom of the jamb, push the bottom of the strike jamb leg in toward the door until the door rests against the stop. If the door is touching the stop at the bottom of the jamb but not at the top of the strike jamb, push the bottom of the hinge jamb toward the other room (see the photo above). Always try to adjust for cross-leg at the bottom of the jamb, not at the top. Keep the jamb head flush with the drywall or it will be difficult to join the miters on the casing. If the jamb is short of the wall at the bot-

If the door isn't lying flat against the door stop then the jamb is cross-legged. Adjust the cross-leg by pushing on the bottom of the jamb, on the strike side or the hinge side, toward the door.

tom of the leg, flatten the drywall with a hammer before installing the casing.

Swing the door open and closed a few more times to check that the door rests flat against the stop. Also check the gaps on the hinge side of the door to be sure nothing has moved. Finally, install shims behind every hinge and behind the lock strike, too. Secure the shims and the jamb by driving two nails through each shim. Before applying the casing to the opposite side of the jamb, score the shims with a knife and break them off behind the drywall. If the drywall is proud of the jamb, take the time to flatten the drywall with a hammer before installing the casing. Split-jamb prehungs are easier to install than standard prehungs because split jambs adjust for uneven wall thicknesses—you never have to fight with the drywall in order to install the casing.

Knock-down prehungs are impossible to shim because the casing is fastened to both sides of the jamb by the manufacturer, which is why these units are equipped with adjusting screws near the top of both jamb legs. Each screw turns a threaded shaft that penetrates the back of the jamb and bears on the trimmer. While the screw head remains flush with the jamb, the threaded

Making Your Own Shims

Most lumberyards sell shims. These wire-bound packages look like miniature bundles of roofing shingles. Builder's shims are meant to be somewhat uniform, and most taper consistently from the thick end, which varies between ⅜ in. and ¼ in., to the thin end, which is usually ⅛ in. to 1/16 in. I use builder's shim's often, but that doesn't mean I like them. They're always too long, they're often too thick at the point end or too narrow at the butt end, and they're always difficult to cut off.

Whenever I'm setting jambs or prehungs, I comb through the job-site scrap pile for 2x6 cutoffs that are free of nails and knots and have tight grain patterns. I cut end-grain shims from these scraps on my portable table saw. It sounds like a lot of work, but it's worth every minute. The shims I cut myself are all 5½ in. long, ¼ in. thick at the butt end, and sharp at the point end. Because they're only ¼ in. thick, I can slip a homemade shim behind most prehung jambs, butt end first, then follow it up with the point of another shim. Stacked tightly atop one another, these shims adjust to fill any odd-size gap between a twisted trimmer and the back of a flat jamb.

Because I cut my shims along the end grain, nails occasionally split them, but these shims remain where they're put and anchor the jamb right where I want it. It's because these shims split easily on the end grain that I like them so much. After shimming a jamb or a prehung door, it only takes a gentle hammer tap to snap off an end-grain shim flush to the jamb. That's a lot easier than scoring a builder's shim over and over and having them still not break off flush.

Cutting shims on a table saw can be dangerous. I've tried to reduce the danger by using a shopmade sled. This template cuts each shim exactly the same size, and neither of my hands are ever near the blade (see the drawing below). My right hand stays firmly on the push block, which keeps it far above and to the right of the spinning blade. I never cut shims from a block smaller than 6 in., so my left hand is always at least that far from the blade. I stand back from the blade and slightly to the left so any chips blow past my right side. I can cut enough shims in a half-hour to set 20 jambs, or almost 30 prehung doors.

Sled for cutting homemade shims

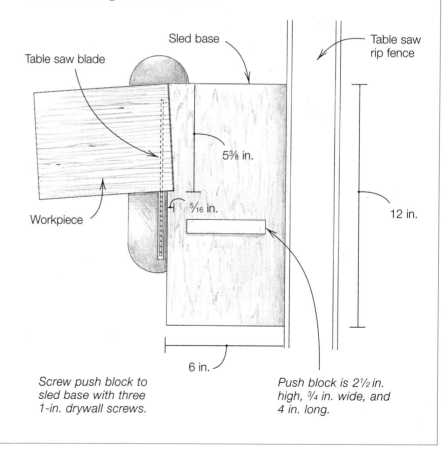

Screw push block to sled base with three 1-in. drywall screws.

Push block is 2½ in. high, ¾ in. wide, and 4 in. long.

shaft acts like a shim between the trimmer and the jamb and anchors the prehung in the opening. For these prehungs, once you've nailed off the casing thoroughly, adjust each screw until you feel the screw begin to resist, which means the shaft has tightened down against the trimmer. Now the weight of the door won't cause the hinge jamb to settle toward the strike jamb.

Installing a pair of hollow-core prehungs Pairs of hollow-core prehungs are awkward to move around—like a pancake poured too big in a skillet—so even when the jamb and doors are in the rough opening, they tend to spill out. To hold the doors and the jamb still, place a shim beneath each door, off center and toward the lock stile. Adjust the shims until the two doors meet evenly across the top and somewhat evenly down the middle. If the floor is out of level, the doors will look something like the pair in the drawing below, in which case one leg has to be lifted from the floor to level out the head and straighten the head gap.

From this point there's little difference between a pair of doors and a single prehung. Use a long level to check that the jamb legs are plumb before driving nails through the casing about 3 in. beneath the miter on each jamb leg. Then move the bottoms of the legs in or out from the center to make fine adjustments in the head gap and strike gap. Nail the bottom of the casing

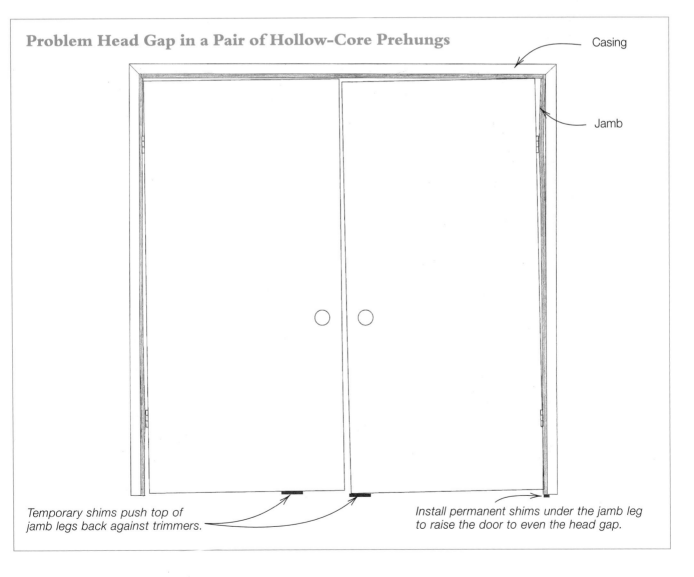

Problem Head Gap in a Pair of Hollow-Core Prehungs

Casing

Jamb

Temporary shims push top of jamb legs back against trimmers.

Install permanent shims under the jamb leg to raise the door to even the head gap.

PREFIT DOORS **81**

to the wall next. The strike gap is normally a little tighter near the top of the doors because the jambs usually deflect a little and have to be pulled back toward the trimmers near the top hinges, just like a single door. Now open one door carefully. Step through the opening to the stop side and drive two nails through the face of the jambs, above and below the top hinges. Those two nails will improve the strike gap by lifting and pulling each door back just a little. Even though the top of the jamb legs are now secured to the trimmers, the bottom of the jamb legs can still be moved to adjust for cross-leg, which on wide openings can be exaggerated. Once the cross-leg is corrected and the doors are flush at the meeting stiles, finish nailing off the casing at each hinge and every 12 in. to 14 in. between the hinges. Case the opposite side, then secure the jamb with four nails, two above and two below each hinge.

Installing a pair of solid-core prehungs

A month or two after finishing an apartment complex, I had to go back to a few units and switch the swing on the bathroom doors because there wasn't enough room in the bathroom for the toilet and the door. Lunch must have interrupted the original installation of one door because there were only two nails shot through the casing into the trimmer on the hinge side of the door and none anywhere else—only the casing on the stop side was nailed to the jamb! But the door must have fit when the painters got to it because they caulked the casing to the wall, and it still fit fine when I found it, too. Which proves a point: It doesn't take many nails to secure a hollow-core prehung door; but solid-core prehungs are different, especially when they come in pairs, and for that reason I use a different method to install pairs of solid-core prehungs.

Shimming the trimmers Solid-core doors cause severe jamb deflection, so it's necessary to use shims behind the hinges and near the top and bottom of each jamb leg. Because shims are so important for solid-core pre-

Preparation is the key to simplifying prehung installation, especially for pairs of heavy doors. Shim the opening before installing the jamb. Tack shims to the studs near the hinge locations.

Measure the opening to be sure the frame will just fit, then slide in the jamb.

Hold the jamb flush with the drywall and tack the legs to the trimmers. Then check that the legs and head are level and plumb.

hungs, I prefer to install pairs without casing—it's easier to place shims and adjust the jamb without the casing in the way. And because shims have to be used anyway, I prefer to start shimming before I slip the jamb into the opening.

Shimming the trimmers is a little like setting the trimmers correctly before drywall is installed. Use shims to plumb the trimmers and narrow down the opening to the outside width of the jamb so that the jamb will fit snugly into the opening, with the legs nearly plumb and the head level, right from the start (see the photos on the facing page). To avoid holding a long level over your head, place it on the floor. Shim the low end until it's level, right where the jamb leg will sit. The preparation pays off when the doors are lifted onto the hinges—they usually fit fairly well and require only minor adjustments.

At first, most pairs of solid-core prehungs will look a little like the drawing shown below, with only minor adjustment problems. The doors weigh so much that the front edges will probably rest on the ground, which tweaks the jamb at the top hinges. To solve the problems, lift each door a little with a prybar and slide in a shim near the center of the opening. Adjust the shims, lifting the weight off the front edge of the doors, until the hinge jambs straighten out and the gap is even across the jamb head and down the meeting stiles. I like to lift the doors a little more than necessary because I

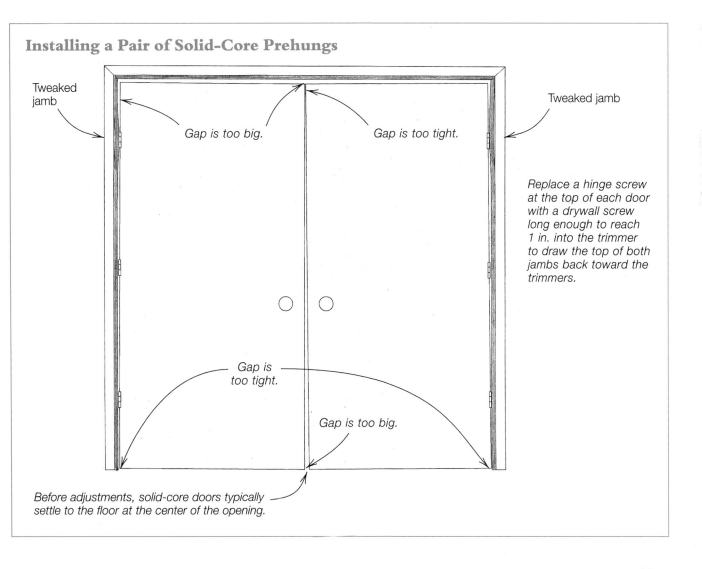

Installing a Pair of Solid-Core Prehungs

Tweaked jamb

Gap is too big.

Gap is too tight.

Tweaked jamb

Replace a hinge screw at the top of each door with a drywall screw long enough to reach 1 in. into the trimmer to draw the top of both jambs back toward the trimmers.

Gap is too tight.

Gap is too big.

Before adjustments, solid-core doors typically settle to the floor at the center of the opening.

PREFIT DOORS 83

If the jamb legs have to be raised off the floor to level the head and even the head gaps, be sure to drive a shim between the jamb and the floor so that the doors won't settle later.

Only screws are strong enough to pull back a tweaked jamb supporting a solid-core door. Remove one of the hinge screws and replace it with a long drywall screw that will reach through the jamb and into the trimmer.

know they'll both settle a little once the shims are removed. If the head is almost even but the tops of the doors aren't flush, then raise one jamb leg just a little, as shown in the photo at left.

Now carefully remove the shim from beneath the active door (the door with the bevel). Swing that door open just enough to get through the opening. Leave the shim beneath the inactive door and secure the inactive door jamb first, using two nails near the top hinge and only one nail near the bottom hinge. Don't drive any more nails near the bottom because the jamb still has to be adjusted for cross-leg.

Because solid-core doors weigh so much, finish nails aren't usually strong enough to secure the top of the hinge jamb. The best practice is to remove one of the hinge screws in the jamb and replace it with a drywall screw (see the bottom photo at left) long enough to pass through the jamb and penetrate the trimmer. I use gold-colored drywall screws because they match the color of most hinge screws and they're available in many different lengths from a variety of suppliers, including most warehouse outlets. To install the screw, run it in slowly, until it's snug. Don't put too much pressure on the screw at first. The pressure of the screw can later be changed to make fine adjustments in the fit of the doors.

Now remove the shim from the inactive door and replace it beneath the active door. Readjust the active door again until it's aligned properly with the inactive door, then secure the active door jamb with three nails and one screw. Swing both doors closed and check the gap at the meeting stiles. If the gap is still too tight at the top of the meeting stiles, tighten each of the long hinge screws just a little more.

Adjusting for cross-leg I correct all cross-leg problems on interior doors before nailing the bottom of the jamb off. If a pair of interior doors is cross-legged after the jamb has been permanently fastened and there isn't a flush bolt in the bottom of the stationary door (which is often the case because flush bolts aren't long enough to reach through carpet and carpet padding), then the only way to adjust the doors is by moving the hinges. Moving hinges should always be the last choice because the result is often unsightly—when the hinges are moved to accommodate cross-leg, the doors are no longer flush with the edge of the jamb.

Because cross-leg can be exaggerated over wide openings, pairs of doors can often be more than ½ in. from flush (as shown in the photo at right). Moving the jamb legs that much might cause the casing to tip at an angle. It's best to try and align the walls before adjusting the jamb legs. Most interior partitions will move a little. If the walls can't be moved because of lightweight concrete or anchors, then there's no choice but to move the jambs. As with all prehungs, once the cross-leg is adjusted, the remaining fastening can be completed quickly. Unlike most prehung doors, on solid-core pairs the casing is applied last.

Installing exterior prehungs Just like on interior doors, the casing is applied last on exterior prehung doors, too. There are also several other similarities between solid-core interior prehungs and exterior prehungs. The trimmers must be shimmed on an exterior opening before the frame is installed, and like exterior jambs (see Chapter 3), the sill must be prepared properly. The sill has to be shimmed level, but it also has to be shimmed to the right elevation so that the door will clear the finished floor covering. And a flashing pan or waterproof barrier has to be installed before the frame.

To position a heavy exterior prehung, first remove the door from the frame, then tack the frame securely but not permanently before hanging the door back on the hinges. Be sure the exterior trim is tight against the framing on the outside of the wall. If the jamb and sill have been properly shimmed and prepared, only minor adjustments are necessary to fit the door. But don't attempt to replace one of the top hinge screws and pull the hinge jamb tighter to the trimmer. For exterior prehungs, that adjustment has to wait until the jamb has been checked for cross-leg, which is the next step.

Because the bottom of the strike jamb cannot be moved farther into the house (the exterior trim should already be tight against the framing), cross-leg often has to be eliminated by moving the bottom of the hinge leg farther outside the house. If the cross-leg is unusually bad, one jamb leg at the top of the frame will have to be pushed toward the outside, which makes casing the jamb a little more difficult. Once the cross-leg has been adjusted, the gaps should be checked again. To permanently secure the unit, replace one of the top hinge screws with a longer screw. Not only will the fit of the

If the cross-leg is really bad, as it is in this pair of doors, use a single jack or sledgehammer to move partition walls into alignment with each other. Protect the drywall with a short piece of 2x4 scrap and smack the wall hard—I mean really hard—near the bottom plate.

door normally require a long hinge screw, but the weight of the door will otherwise pull on the jamb and eventually the door will rub on the strike jamb or the sill.

Exterior sills rarely require elaborate fastening because of the silicone water barrier applied before the frame is installed. One or two screws are usually sufficient to secure a sill. If the sill is aluminum, replace one of the sill screws with a screw long enough to reach into the subfloor or concrete. If the unit has an oak sill, counterbore for an oak plug before installing the screw. Concrete screws, like Tap-con screws, are useful for this purpose because no anchor is required—the hole can be drilled and the screw installed without removing the unit from the opening.

Installing a pair of exterior prehungs Pairs of exterior prehungs are also similar to pairs of interior solid-core prehungs, except for one thing—cross-leg and flush bolts. Follow the same preparatory steps I described for a pair of solid-core prehungs: Shim the trimmers plumb and measure the opening so that the frame just fits; prepare the sill in the same manner as you would a single exterior prehung. Once the doors are swinging and the gaps are adjusted, check that the doors are flush at the meeting stiles. Adjust for cross-leg by first moving the bottom jamb legs. Unfortunately, the top of the jamb may have to be moved as well.

Most pairs of exterior prehungs include astragals with flush bolts. The top flush bolt hole and strike are normally installed, but the bottom hole is rarely drilled and the strike is supplied loose so the jambs can be adjusted. Remember to remove the strike before in-

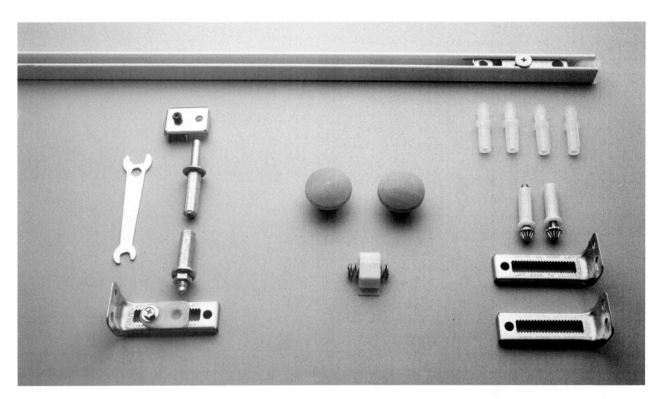

Although there are a lot of parts in a bifold hardware bag, once the parts are separated they all make sense. The jamb bracket and bottom pivot are on the lower left (the pivot is pointing down where it seats in the bracket). Above the bottom pivot is the top pivot and pivot guide, which mounts directly in the track (shown at top). On the right is the hardware for a pair of bifold doors.

stalling the frame, and try to remove all the cross-leg before locating the bottom flush bolt hole. A slight amount of pressure can be applied to the bottom flush bolt, which can alleviate a small degree of cross-leg or warp in a door, however, applying pressure to the flush bolts is risky and I only do it when there are no other alternatives. Too much pressure on a flush bolt can split the bottom of a door, even a little pressure on a flush bolt can break a finger nail and cause a nasty call-back. The flush bolt holes should be located so that the flush bolts operate smoothly, without any resistance. (See p. 143 for information on how to use a flush bolt locator.)

PREFIT BIFOLD DOORS

The subject of prehung doors had to include exterior doors, which led me right outside. Now I'll go back inside for more on prefit doors, starting with bifolds. Bifold doors are the perfect answer for closets or laundry areas, especially where swinging doors consume too much space from hallways and bedrooms. Bipass doors, also called sliding doors, were once the only choice for closets, but bifold doors have a significant advantage over sliding doors: Bifolds (shown in the photo at right) fold and pivot back to the jamb, and but for the thickness of the folded doors (about 3 in.), leave an unobstructed opening. Bipass doors, on the other hand, slide back over each other, so half the opening is always covered by the width of one door.

Bifolds are packaged with all the doors and necessary hardware, including the track. Stanley/Acme 2400 series hardware (shown in the drawing on p. 88) and similar products are designed for doors weighing up to 30 lbs. The 2400 series is the most common type of hardware packaged with a set of prefit bifold doors. The 2400 series overhead track is really only a pathway for the top roller guide on the leading door. All of the weight of the doors is supported by the top and bottom pivots in the rear doors. Many other types of bifold hardware are available, and I recommend choosing carefully. Often the hardware selected for an opening isn't heavy enough for the doors or the aggressiveness with which they are operated, which is one reason why folding doors have lost some appeal.

Sizing the jamb and the doors Everything is usually included with a package of bifold doors, except the jamb and casing, and they are the first things to install. I use standard jamb stock, matching the material I've used for the other doors on the site. The 2400 series bifold hardware requires an additional 1⅝ in. in height, but because jambs are typically ½ in. taller than stan-

Folded open against the jamb, a bifold door only takes up about 3 in. of space from the door opening. But remember, a cabinet installed inside the closet must be placed so that the drawers and doors clear the open bifold door. The best location for a bifold knob is on the front stile of the pivot door.

PREFIT DOORS 87

Prefit Bifold Installation

Stanley/Acme 2400 Series

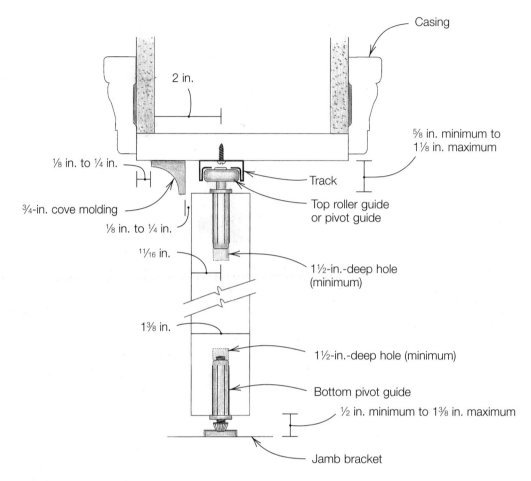

Four-door bifold

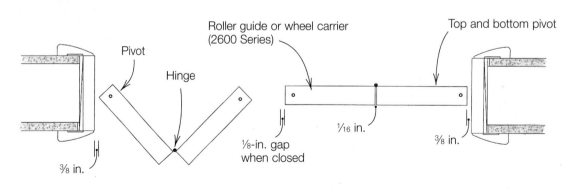

88 CHAPTER FOUR

dard 6-ft. 8-in. doors, and because manufacturers commonly cut bifold doors 1 in. shorter than standard 6-ft. 8-in. doors, there's usually just enough room for the bifold track and the bottom pivot in an opening with a hard-surface finished floor or vinyl floor. However, there isn't enough room for carpeting.

Though it might be tempting to raise the jamb and avoid cutting the doors, you can't. The header is normally set at 82 in. from the floor, which means the jamb can't go any higher; and besides, the top of the jamb and the casing have to line up with adjacent jambs, which prevents building a taller jamb for thicker floor coverings. For carpeting, the doors have to be cut and the jamb brackets raised off the floor. The bottom and top rails on many bifold doors are made from 1-in. MDF. If more than ½ in. has to be cut off the doors (try not to cut more than ¼ in. from the top rail and the bottom rail), then the doors have to be cut and plugged (see the sidebar at right).

The width of a bifold opening should be a standard door size. If the plans call for a 4-ft. pair of bifolds (two 2-ft. doors), then the finished opening should measure 48 in. Manufacturers undersize bifold doors in width, too. Each door in a bifold set is typically planed down ¼ in. by the factory, which allows room for the doors to pivot and clear the jamb.

Installing the Jamb and the Hardware

Assemble bifold jambs on the floor and attach the casing, too. Like all prehungs, it's easier to set the jamb if the casing is on one side. Before lifting the frame into the opening, lay a long level from trimmer to trimmer and level the floor where the jamb legs will sit. That eliminates having to hold a long level over your head. Set the jamb centered in the opening. Plumb the legs and get the head straight while nailing the casing to the wall, then case the opposite side, too. Afterwards, nail through the face of the jamb into the trimmers—sometimes people change their minds and decide on swinging doors, so set the jamb well.

Installing the track is the next step, but first the trim technique has to be decided. There are several popular ways to trim bifold doors. I prefer to use ¾-in. cove molding all the way around the opening. The cove molding hides the track and the wide ⅜-in. gap between

Cutting and Plugging Bifold Doors

If bifold doors haven't been ordered correctly to fit the finished opening height, they often have to be cut and plugged. To do this, first cut the door off at the necessary height. Save the lock rail that falls off on the floor. Strip the hardboard or veneer off both sides with a sharp chisel or sander. If I'm cutting a quantity of doors, I run the lock rails through my table saw.

Spread glue on the clean rail and reinsert it in the bottom of the door. Clamp the door for an hour, but wait overnight before inserting the pivots, because the glue won't yet be strong enough to support the door. A few doors can be clamped together, which reduces waiting time. Just place waxed paper between the doors so they don't stick together. By using the old lock rail, new pivot holes don't have to be located.

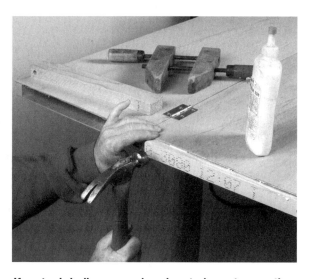

If a stock hollow-core door has to be cut more than 1 in., then the door will have to be plugged. Save the bottom stile, scrape off the hardboard, and sand down the old glue, then spread fresh glue and reinsert the stile in the bottom of the door. A few clamps and an hour or two later, and the door's as good as new.

the doors and the jamb (see the drawing on p. 88). Others would rather not see trim around a pair of bifolds, though some type of top trim is required to cover the bifold track. Stanley Hardware illustrates their installation instructions with a piece of ½-in. molding, like base shoe, installed flat along the ½-in. track. I've also seen installations where a piece of ¾-in. by 1½-in. trim is nailed flat across the top of the jamb. The head casing is then dropped to cover most of the ¾-in. trim, leaving a standard 3/16-in. reveal. This looks good, because the track is hidden behind the trim and the trim isn't noticeable in the opening. But before dropping the casing ½ in., be sure there are no adjacent doors or windows, because all the head casings in a home should be in a level line (except where floor elevations change).

For cove molding, the center of the top track has to be set back from the face of the jamb about 2 in. (see the drawing on p. 88). Spread a pair of scribes 2 in. apart and scribe a line along the jamb head. While the scribes are still spread 2 in., scribe a line for the jamb bracket on the jamb leg, up to about 6 in. from the floor (2 in. is the center of the track, which is the center of the doors, which is also the center of the jamb bracket). Position the track on the jamb head by aligning the screw holes in the track with the pencil line. Fasten the track with the short screws that came in the parts bag. Use the same technique to install the jamb bracket, but use the long screws so they'll reach the trimmers. Don't tighten the screws too much or they'll pull the jamb.

For hard-surface floors, like hardwood, tile, and stone, I install the jamb bracket tight to the floor; for carpeted floors I first install a ¾-in. block of wood, the width and length of the bracket. The block of wood supports the weight of the doors, and the carpet can be cut around the block, hiding it entirely. Once the track and brackets are on the jamb, install the pivots and roller guides in the doors.

To eliminate confusion when installing the hardware in the doors, stand each pair of doors up and lean them against or near the opening. Insert a top pivot in the top rear hole of each rear door. These are the pivot doors. Install a top roller guide in the top front hole of each front door. These are the guide doors. The bottom pivot has to be installed in the bottom rear hole of the rear pivot door, so lay the doors back down and use a hammer and a block of wood to tap the pivots into the holes. If the pivots fit too tightly in the holes, ream out the holes just a little with a slightly larger drill bit—forcing a pivot might split a door.

Installing the Doors and Trim

All bifold doors are installed the same way. Fold one pair of doors in half, so they're easier to handle, and hold them next to the jamb, right beside the jamb bracket. Lift the doors and insert the top pivot into the top pivot guide at the end of the track. Keep lifting the doors while the spring-loaded top pivot depresses in the pivot guide and the bottom pivot clears the jamb bracket, then lower the doors until the bottom pivot engages the jamb bracket. The spring-loaded top roller guide can be popped into the track anytime after the doors are swinging.

Bifold doors can be adjusted in three ways. Turning the bottom pivot counter-clockwise will raise doors clear of the floor and close the gap between the top of the doors and the jamb head. Be sure to leave at least ⅞ in. between the tops of the doors and the jamb head, or the cove molding won't fit. Turning the bottom pivot clockwise will lower the doors. Use these adjustments to align the tops of both doors. The meeting stiles can be adjusted either by moving the bottom pivot to a new location in the jamb bracket or by moving the top pivot guide in the track. Moving the bottom pivot is easy—just lift the weight of the doors off the bracket, then relocate the pivot. But move the top pivot carefully. Only loosen the top pivot screw enough to pry the guide to a new location. If the screw is loosened too much, the guide and the doors will slide across the track and the doors will fall out.

Knob placement on folding doors is ambiguous. Some installation hardware suggests that the knob be located in the center of the leading, or guide, door. But there's no support in the center of a hollow-core door, and the knob always works loose. Besides, the doors can't be pulled closed and pulled open with only one hand if the knob is in that location. Accidental experimentation has convinced me that the best one-handed location for a bifold knob is in the front stile of the pivot door (see the photo on p. 87).

Bifold Hardware

Major manufacturers of bifold and bipass units offer several lines of hardware. Never use the least expensive products; always step up to the next line if there's any doubt about the weight of the doors, the frequency that they will be operated, or the strength of the anticipated operator. Avoid hardware with too many plastic parts. Stanley's specifications for its bifold hardware illustrates my point.

While the 2400 series is the most common set of hardware installed in residential construction for door panels weighing up to 30 lb. each, Stanley's next line, the 2600 series, is meant for door panels weighing up to 50 lbs. According to Stanley, the series is "designed for high frequency installations such as those in hotels/motels, hospitals and schools." I recommend the 2600 series for many openings, especially a child's bedroom and particularly a boy's bedroom.

The quality of the 2600 series hardware is improved over the 2400 series: The track is heavier, extruded aluminum, the front roller guide is a three-wheel, spring-loaded hanger, and the bottom pivot is made completely from metal—the entire package is built to last. The initial cost is more, but the long-term savings is substantial.

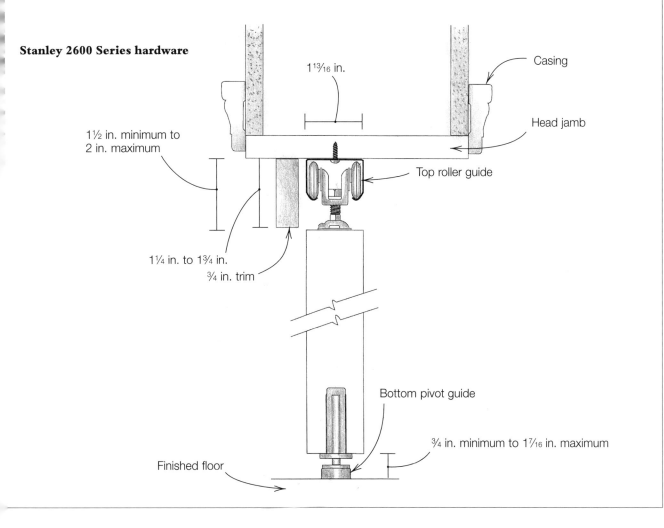

Stanley 2600 Series hardware

Bipass Doors with Wood Fascia

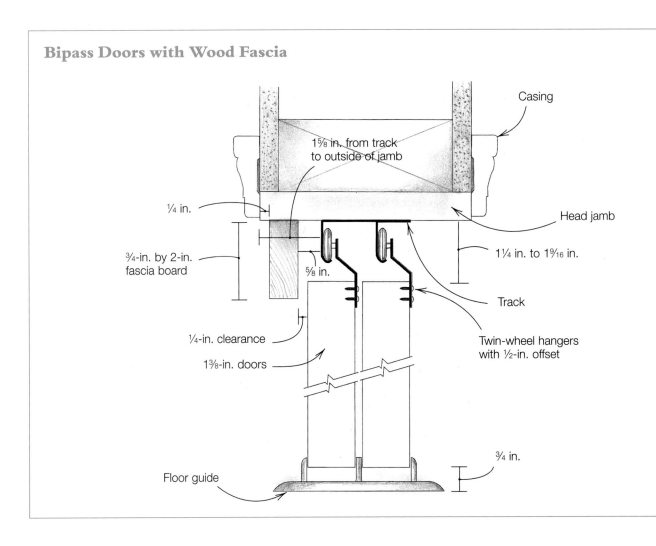

To correct any cross-leg in the jamb and to keep the doors flush to each other at the center of the opening, a pair of aligners is included with every four-door set of bifolds. Attach the aligners to the backs of the leading doors, about 12 in. from the floor. Space the aligners so that they just clear one another, and they'll almost snap together as the doors are closed. Also, if the opening is for a four-door set, snap the snugger into the track between the doors, so that the doors will close "snug" against each other tightly and stay that way, rather than hinging open a little on their own. Now cut and nail on the cove molding.

PREFIT BIPASS DOORS

I prefer using bifolds in most closets because, with the doors folded back, they allow a greater degree of access. However, there are cases where bifold doors are not the answer. Bifold doors fold open against the jamb and take up about 6 in. of the opening. A cabinet with drawers or doors inside a bifold opening has to be positioned a considerable distance past the line of the jamb so that the doors and drawers will clear the bifolds when open. Sliding bipass doors are sometimes the only alternative, particularly when a laundry closet must be squeezed into a space less than 5 ft. Sliding doors allow clear access beside the jamb, though only to one side of the opening at a time.

Sizing the Jamb and Doors

Similar to bifold doors, bipass doors have a track and wheels at the head of the jamb, but they have a guide on the floor, too, so the space required for the top track and bottom guide has to be calculated when ordering the doors. The most common hardware used on bipass doors is similar to the Stanley/Acme series 8800 in the drawing on the facing page, though, like bifold hardware, lighter-duty and heavier-duty track and wheels are also available (see Chapter 7 for more on these types of hardware). However, I don't recommend using single-wheel hangers. Twin-wheel hangers operate smoother and last longer.

Packaged bipass doors are also designed to fit into standard 6-ft. 8-in. jambs that measure 80½ in. from the floor to the jamb head. Manufacturers typically cut down bipass doors from 6 ft. 8 in. to 6 ft. 6½ in. to accommodate the necessary clearance for the track, wheels, and floor guide. The 8800 series requires a minimum 1¼-in. clearance between the top of the doors and the jamb head. An additional ¾-in. clearance is needed for the floor guide. Like bifold doors, bipass doors often have to be cut for carpet. I generally allow 1⅜ in. for a door to clear carpet, and I add another ¼ in. for the floor guide.

Like bifold doors, stock lock rails in bipass doors are manufactured from 1-in. MDF and are installed with the grain of the MDF parallel to the face of the door so that bipass hardware screws will hold well. Not much can be cut off the top of a bipass door without jeopardizing the strength of the door, particularly since the wheels and the weight of the door hang from the top lock rail. If the doors are not ordered correctly, then they often have to be cut and plugged.

Unlike the width of a bifold opening, which should be a standard door size, the width of a bipass opening should be 1 in. less than a standard door size, to allow room for the doors to overlap. If the plan calls for two 36-in. doors, then the opening should measure 71 in. wide, not 72 in. When I install bipass doors in an existing jamb that isn't 1 in. undersized, I also install a pair of bumper jambs—two jamb legs ripped down to approximately 3½ in. so that they fit on the face of the existing jambs. Once bumper jambs are painted or stained, they add one more level or step in the molding detail. Bumper jambs are also useful on openings that are wrapped with drywall. It's better to have doors banging into wood or MDF bumper jambs than into drywall or metal corner beading. I usually wait until after the upper track is in before installing the bumper jambs, but I've seen other carpenters install bumper jambs first. There's no right way.

Installing the Hardware

Bipass door technology, like the wheel itself, is pretty simple. For a two-door set there's four wheels. Mount the wheels on the back of the doors, about 2 in. in from the edge. If the wheels are too close to the edge of the doors, they'll reach the end of the track before the door hits the jamb. Most wheels have a crease or bend just above the screw locations. Place the bend near the top edge of the door. One of the screw locations is a round hole, and that screw can be tightened securely. The other screw location is an elongated slot and that screw should just be snug so that later adjustments are easier.

On some models, the fascia—a flat trim piece that covers the track—is part of the track. If the track includes the fascia, cut the track and fascia carefully, because the cut is exposed. Often the fascia isn't supplied with the hardware and instead a piece of 1x trim is used.

If the bipass track includes fascia, then install the entire assembly just back from the face of the jamb, because the fascia looks better if it's not flush with the jamb. If the fascia is not included, then the track has to be installed far enough back from the face of the jamb to allow for the thickness of the fascia (see the drawing on the facing page). I usually read all installation instructions, then mock up the product to be sure I have everything right, especially the track and wheels. Wheels are manufactured with different offsets, and the offset affects the position of the doors in relation to the face of the jamb. I add an extra ¼ in. to allow plenty of clearance between the doors and the back of the fascia—there's no reason for the fascia to be really close to the doors.

To locate the position of the track on the jamb head, spread a pair of scribes 1⅝ in. apart and scribe a line down the jamb head. That line is the front edge of the 8800 series track, anticipating a ¾-in. piece of fascia. Because the wheels hang from the track, attach the track with the long screws supplied in the parts bag. These screws will penetrate the jamb and the header above it.

Mirror Bipass Doors

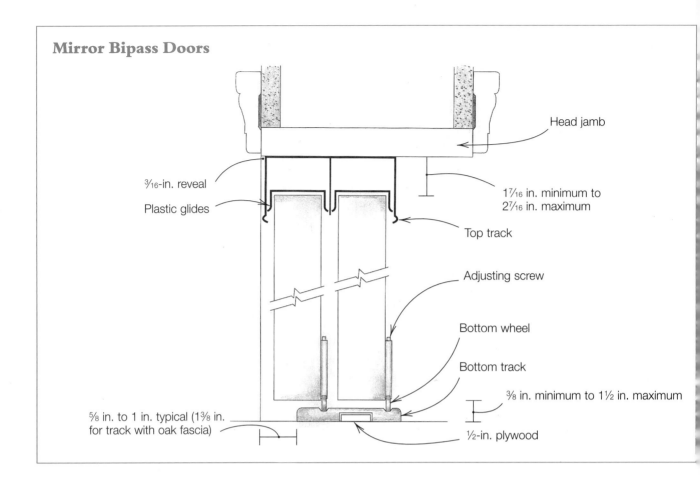

Hanging and Adjusting the Doors

Unlike bifold doors, which lift straight up into the pivot guides, a bipass door must be lifted at an angle. Start with the rear door and tilt the door so that both wheels clear the back edge of the track, then lower the door onto the wheels and allow it to hang straight to the ground. Before adjusting either door, get the correct door against each jamb leg—there's a right and a wrong way for bipass doors to close. The vertical gap between a pair of sliding doors shouldn't be visible from the entrance door to a room; rather, the doors should be positioned so that the gap is pointing away from the bedroom door.

With the doors hanging against the correct jamb legs, stand inside the opening with a flat-blade screwdriver. Insert the blade of the screwdriver into the hanger slot and turn the screwdriver to adjust the fit of the doors against the jamb. Either lower the front wheel or the rear wheel until the edge of the door lies flat against the jamb. Once the doors are adjusted, tighten the second screw holding the wheel to the door, then nail on the fascia and attach the floor guide.

When bipass doors are installed over a hard-surface floor, the floor guide can be mounted directly to the floor. With both doors open, insert the guide between the doors and hold the front of the guide securely to the floor with one hand. Slide one door closed so that both doors are against a jamb leg. Still kneeling on the floor, look up and down the leading edge of the front door (the door closest to the front edge of the jamb) and move the floor guide and the door until the edge of the door is parallel with the edge of the jamb. Use a sharp pencil to trace the footprint of the floor guide, then open the doors and install the guide.

When bipass doors are installed over a carpeted floor, which is often the case, install a block of wood to lift the base of the guide through the pile of the carpet. Use a 1 in. x 1 in. x 5 in. (net size) pine block (which is the same size as the floor guide) beneath a floor guide in a carpeted room. (Hagar also manufactures a metal carpet raiser, part no. 9354.) The same technique is used to locate the position of the block, only the guide and the block are held together. Drywall screws work well to secure the block and the guide to a wood subfloor. On concrete floors, use plastic concrete anchors and glue, too. Locate the footprint of the block, secure it, then attach the guide to the block with short screws. Hopefully the carpet installer will remove the short screws and leave the block, cutting the pad and carpeting around the block. Unfortunately, carpet installers have a habit of removing the blocks, and anything else they might have to cut around.

It's difficult to replace a block after the carpet has been laid. The carpet has to be cut and a portion of the padding removed so that the block can sit flat on the concrete or subfloor. Rather than cutting out a rectangle, make one slit in the carpet, so the block will just fit in. Leave the carpet tight to the block so that the carpet installers have to come back to trim the carpet. I always hope they'll learn to leave the blocks in place, but there must be a lot of carpet installers—I have yet to meet the ones who have learned.

MIRROR BIPASS DOORS

Mirror bipass doors are normally installed after a home has been painted, which means the jambs are set and finished and it's a much cleaner job. You don't even need a circular saw—if the doors are too tall for the opening, just return them. Mirror doors have to be ordered properly, but there's usually a large latitude between minimum and maximum height. Most 6-ft. 8-in. units are adjustable from $79\frac{3}{4}$ in. to $80\frac{3}{4}$ in., which means that mirror doors should be installed before carpeting. Stock sizes are not made short enough to fit into a carpeted opening, and custom sizes are much more expensive.

Unlike bifold and bipass doors that have a continuous top track but no continuous bottom track, most mirror doors also have a continuous bottom track. And rather than the wheels hanging from the upper track, the wheels ride in the bottom track, like an exterior sliding door, and slide through plastic guides in the upper track. Because a mirror door track is prefinished and includes fascia trim, the upper track can be installed just behind the face of the jamb. Short screws are more than adequate to secure the upper track, as the wheels and the weight of the doors ride on the lower track. The location of the top track is simple, but the bottom track on mirror doors isn't always installed flush with the jamb. Different manufacturers require different setbacks for the bottom track, though 1 in. from the face of the jamb is common. Check the instructions supplied with the unit and mark each jamb leg for the proper setback, then locate the track in the opening.

Mirror doors often come with a narrow strip of $\frac{1}{2}$-in. plywood that helps secure the track to the floor. A hollow in the center of the track slips over the plywood strip, and short screws, supplied with the track, are used to fasten the track to the plywood. A handful of short common nails is included in the package to fasten the plywood strip to a wood subfloor. A handful of short concrete nails is also included for installing the track on a concrete slab. I recommend throwing the concrete nails away. They rarely work and only make a mess out of the concrete. Instead, to install the track over concrete, lay the track in the opening and run a $\frac{1}{8}$-in. masonry drill bit through three or four screw holes. Remove the track and use a larger bit to install plastic anchors in those holes, then screw the track and the plywood to the concrete. Fill in the remaining holes in the track with short screws.

Hanging the doors is the easiest part. Lift the doors up into the channel of the top track, then lower them onto the bottom track. Before sliding the doors, lower the wheels because they are usually retracted for shipping and the doors might scratch or scuff the bottom track. Slide each door up against the appropriate jamb and adjust the door parallel to the jamb by lowering the back wheel or raising the front wheel. Be sure to lower the wheels far enough so that the tops of the doors are securely contained in the upper channel.

Chapter 5
HANGING DOORS

SCRIBING A DOOR

CUTTING THE TOP AND BOTTOM

PREPARING THE HINGE STILE

PREPARING THE LOCK STILE

HANGING THE DOOR

As carpenters, we all share the desire to construct with our hands something we've only imagined. Creating what we imagine is possible by following a simple series of practiced steps. Hanging doors is a simple series of steps, too, and in this chapter I'll explain the process and steps I use to hang single doors.

Everyone in the trades agrees that a door should have an even margin, a perfectly straight line—about the thickness of a nickel—up the hinge stile, across the head, and down the lock stile. That's the way doors are supposed to fit—not too tight, not too loose, just right. But there's no agreement on how to hang doors. Some carpenters first cut the door to fit snug in the opening, then plane and plane until the margins are perfect. Others recommend taking careful measurements, then testing and fitting and planing. But both of these methods are time-consuming and difficult. Even though the results are often pleasing, repeated trips to the jamb with the door are exhausting, especially with heavy doors. I avoid all that hard work by carefully scribing each door so that it fits the first time up. If there are any tricks to hanging doors, scribing is the most important one.

This homemade door hook holds the door tight against the jamb for scribing.

SCRIBING A DOOR

Before you can scribe a door to a new opening, be sure it's not taller than the jamb. If it is, cut enough off the bottom so that the top of the door isn't above the head of the jamb. Never cut too much off the top of a door. Most doors, like raised-panel and molded-panel types, have 4-in. top rails and 4-in. stiles. If too much is taken off the top of the door, then the rails and stiles won't match. Also, very little should be cut off a flush door because the top rail is often only 1 in. thick (see Chapter 1 for more on this).

Set the door down at the opening on two long shims. I cover my shims with masking tape because the friction of the tape stops the doors from slipping. The tape also saves my shims from being thrown away or lost. I use a homemade door hook (see the photo above and the sidebar on p. 98) to pull the door against the jamb and hold it there, but before doing anything else, I always pencil an X at the top of the door on the hinge side. If the door is dark or prefinished, stick a piece of tape on the top corner of the hinge stile and mark an X there. That simple bold X will prevent the mistake of cutting the top of the door instead of the bottom; it will prevent hinging the door on the lock stile rather than on the hinge stile; it will prevent mortising the hinges out the stop side of the door rather than on the front of the door; and it will prevent laying out the hinges from the

As shims and a door hook hold the door steady, carefully center the door in the jamb. Measure from the sticking on the stiles to the jamb and from the sticking on the top rail to the jamb.

bottom of the door rather than from the top of the door. In short, if you make that X every time—right away— you won't have a useless, funny-looking door leaning against some wall, with a lock bore 36 in. down from the top rail, instead of 36 in. up from the bottom rail.

HANGING DOORS 97

A Homemade Door Hook

In order to scribe a door, the door has to be held tightly against the jamb. The door can't move at all, and that's the purpose of a door hook. Made from two hooks connected by a length of bicycle inner tubing, a door hook slips over the top of the door, stretches under the head jamb, then hooks over the other side of the jamb. With one hook at the top and two shims at the bottom, any door will stay still for scribing.

In the past I've used plate strapping to make door hooks, but steel is heavy and I get tired of heavy things in my tool pouch, so I use aluminum now. Aluminum is also easier to bend and drill than steel. The hook that goes around the door should be 1¾ in. deep to hold a standard door securely. My 1¾-in. hooks have always worked fine on 1⅜-in. doors, too, because smaller doors are lighter and easier to hold. For doors larger than 1¾ in. I use a door hook that's paid by the hour and helps me carry the door, too!

The hook that wraps around the jamb should have a serrated edge so that it grabs the jamb securely, but the edges shouldn't be sharp enough to scratch the jamb or paint. I use a grinder to cut small serrations in the edge of the hook, then I file and sand the cuts until they're smooth and round. A piece of 1¼-in.-wide by 8-in.-long bicycle inner tubing works well between the two hooks. The distance between the two hooks shouldn't be more than 3½ in. because most interior jambs are between 4½ in. and 4¾ in. wide.

The hook for the jamb side should have a hole for the rubber, so that no fasteners rub against the jamb. Lay out the hole slightly wider than the rubber. Drill a hole at each end of the layout, then cut the middle out with a small jigsaw and a metal-cutting blade. Thread the inner tubing through the hole.

I use a different method to attach the inner tubing to the hook on the door. Rather than cutting a hole, I cut a backing strip of aluminum as long as the width of the hook. The width of the backing strip doesn't matter, but ¾ in. works fine. Drill a ¼-in. hole through each end of the backing strip and a matching hole through each end of the hook on the door. Loosely join the two pieces with 3/16-in. by ½-in. hex-head bolts, leaving ample room between the strip and the body of the hook to thread the two free ends of inner tubing. Pull the inner tubing through the metal sandwich until the two hooks are about 3½ in. apart, from the part that hooks on the door to the part that hooks over the jamb. The excess inner tubing sticks out the top of the door hook. When I work in a house with 6-in. jambs, I adjust the length of the rubber for the wider jambs.

The knob is the last touch. Drill a hole through the back of the jamb hook to mount the knob. The knob makes it easy to stretch the inner tubing across the jamb.

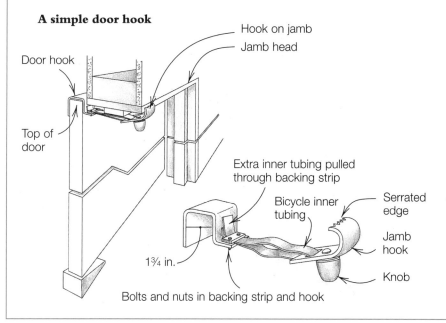

A simple door hook

Scribes are essential and cheap. But use a mechanical pencil in the scribe, and you won't have to pull out a utility knife every time the lead breaks.

A common lead pencil, held flat against a jamb, scribes a line ⅛ in. from the jamb.

Centering the Door

With the door shimmed off the floor and hooked against the jamb, raise, lower, and shift the door side to side until it's perfectly centered in the opening. I carry a prybar in my tool belt to adjust the door incrementally until it's positioned perfectly. To check the position of the door, measure from the head of the jamb to the bottom of the top lock rail at both sides of the door. Those measurements should be the same before scribing any pencil lines.

Try to center the lock stile and hinge stile in the opening, too, so that an equal amount of material can later be cut off each side of the door. Measure from the jamb to the sticking or molding on both the hinge stile and the lock stile—at the top of the door and the bottom of the door—to ensure that the door is centered in the opening. Sometimes jambs are undersized and a lot of wood has to be planed off the stiles, but be sure the lock stile will be at least 3⅞ in. wide before scribing any pencil lines. Some lockset latches, like Schlage dead bolts, are 3⅝ in. deep—so a minimum 3⅞-in. lock stile is vital!

Scribing Four Sides

I use a set of scribes manufactured by General Tool Company that cost about $2 (see the left photo above). Any pencil fits them, but I prefer to use a mechanical lead pencil. If the lead breaks on a mechanical pencil it's easy to fix with one hand, which is nice when you're on a ladder, at the top of an 8-ft. door, with one hand braced against a wall and a flashlight in your mouth.

A regular pencil held flat against the rabbet (broken or short pencils work best for this) will scribe a line on the door about 3⁄16 in. from the jamb (see the right photo above) which is how far I spread my scribes—most of the time. I scribe a little more off a 1¾-in. door than I do a 1⅜-in. door because the thickness of the door de-

HANGING DOORS **99**

The Door Bevel

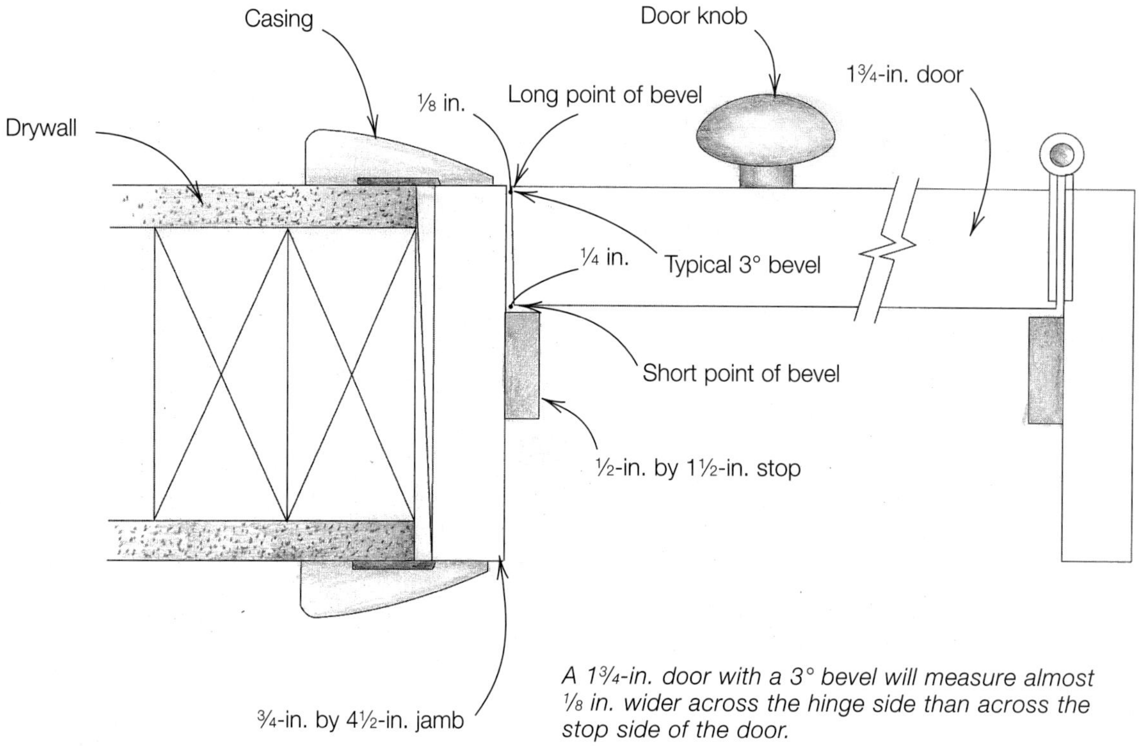

Labels on diagram:
- Drywall
- Casing
- ⅛ in.
- Long point of bevel
- Door knob
- 1¾-in. door
- ¼ in.
- Typical 3° bevel
- Short point of bevel
- ½-in. by 1½-in. stop
- ¾-in. by 4½-in. jamb

A 1¾-in. door with a 3° bevel will measure almost ⅛ in. wider across the hinge side than across the stop side of the door.

termines the width of the bevel, and width of the bevel affects the size of the door. With a typical 3 degree bevel, a 1¾-in.-thick door will be almost ⅛ in. wider on the hinge side (or long point of the bevel) than on the stop side (the short point of the bevel), as shown in the drawing above; a 1⅜-in.-door will be slightly less, about ¹⁄₁₆ in. wider on the hinge side. But don't think that I've figured out the exact relationship between the thickness of the door, the degree of the bevel, and the spread of my scribes. Trial and error has convinced me that—for 1¾-in. doors—spreading my scribes ³⁄₁₆ in. apart works and results in a ⅛-in. margin. Trial and error will also determine the exact setting for your scribes: The degree of your bevel and your planing technique (do you plane to the line or leave the line?) will establish the spread of your scribes.

On a 1¾-in. door, like the one shown in the left photo on p. 99, spread the scribes ³⁄₁₆ in. apart for the hinge side and the strike side (bevel all doors on both edges so that they'll never rub or bind on the jamb and so that the leaves of the hinges will never touch). Hold the scribes perpendicular to the jamb. Don't press against the jamb too hard or the scribes might accidentally close. Trace a nice sharp line on the door. If the grain in an area interferes with the lead, try scribing that area in the opposite direction.

Squeeze the scribes completely closed to trace the head of the jamb, because the head of the door doesn't get a bevel. Adjustable scribes are also handy for scribing the bottom of the door. The scribes can be spread the thickness of the floor covering, which varies from 1⅜ in. for carpet to ¼ in. for vinyl. Try to find out what the floor covering will be before scribing a door. It's much faster to scribe and cut the door now, and eliminate having to take it off the jamb later.

Before removing the door from the jamb, check that all four sides of the door have complete scribe lines. Be sure there's an X on the hinge stile of the door. Don't get in a hurry during the scribing stage. Careful attention pays off in a better fit and less work.

Locating Hinges and Locksets

The hinge and lockset layout can be marked by sighting across the hinge and lock mortises on the jamb while the door is standing in the opening, but it's safer and more accurate to use a tape measure. Pull a tape measure down the hinge side of the jamb and hold it tight with one finger against the jamb, just below the top hinge. The end of the tape should hit the top of the jamb. Think of the tape measure as a door, and slide the top down from the head of the jamb a little more than ¹⁄₁₆ in., about the thickness of a nickel. You can measure the size of the gap as the tape measure drops below the top of the first hinge. The measurement at each hinge will be the exact location of the hinge on the door (see the photo at right). Write those measurements lightly on the face of the door, near each hinge location, or on a scrap of paper.

Measure the location of the lockset, too. Measure down the strike side of the jamb to the center of the lockset strike. Subtract ⅛ in. from that measurement for the head gap and write the result down on the door, near the location of the lockset. If these measurements are made carefully, then the door will fit and the lock will latch. On raised-panel doors with lock rails, I like to center the lockset in the lock rail. I'd rather move the existing strike mortise on the jamb than have the lockset look out of center in the door. For this example then, I'll drill the door where I want and fill the jamb later.

To measure the hinge layout, imagine the tape measure is the door. Hold the tape measure hook down from the head of the jamb about the thickness of a nickel to simulate proper head gap.

HANGING DOORS 101

Door Stands and Benches

Tools are important in door hanging. And one of the most useful doesn't cost more than a few dollars: a door stand. A door stand is a simple clamping device that rests on the ground and holds a door securely on edge while you plane the stiles, mortise for hinges, and drill for locks. A door bench, on the other hand, not only holds a door on edge but it also holds all your tools up off the ground, within easy reach. A door bench is also a smooth carpeted surface—like a big sawhorse—and holds a door flat so that the top and bottom can be cut.

There are many manufactured door stands, like The Rok Buk and The Door Buk, both made by IDMM System. These metal stands are lightweight and durable. The Rok Buk is on wheels and doubles as a door dolly. The Wedge, another manufactured stand, is also simple, durable, and light.

However, homemade door stands are even more common than manufactured stands (one type is shown in the drawing below). I even know one hanger who uses a 3-ft. piece of 4x6 on edge, with notches cut in the edge for different door thicknesses.

If you only install a few doors a year, a homemade door stand is perfect for holding a door on edge. But you'll need a sturdy pair of sawhorses because a door has to be laid flat, too, so that the top and bottom can be cut. If you hang more than a few doors a year, and you're tired of bending over to pick your tools up off the ground, then a door bench or door box is the way to go.

The door bench shown in the drawing on the facing page is one I made. The dimensions and details have been refined through the years and include ideas from boxes I've seen other hangers use. Door bench design is a creative art, but there are specific reasons for most details. The bench has to be narrow enough to fit through a 2-ft. 4-in. doorway, and it has to be long enough to support an 8-ft. door. My bench is 22 in. wide and 68 in. long.

The height of a door bench depends on the height of the carpenter. I'm 5 ft. 7 in. tall and my bench is 32 in. high, which is high enough to work at comfortably, but still short enough so that when I step inside the box, to carry it from one location to another, the legs of the box clear the ground by a safe margin. The height also allows me to climb stairs while walking *inside* the bench, without having to lift the bench over my head!

Because one of the primary purposes of a door bench is to

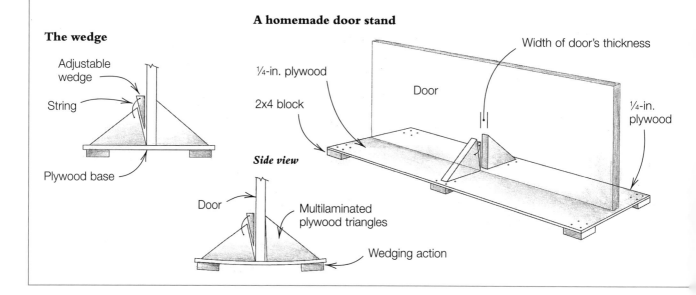

The wedge

A homemade door stand

102 CHAPTER FIVE

support doors on edge while they're being machined, a door box should have adjustable rungs for varying widths of doors. The best layout I've found is to have a rung every 5 in., measuring from the ground up. The legs can be made from scraps of 2x4, with 1¼-in. rung holes drilled with a spade bit or hole saw.

I spent extra time building my door bench because I wanted it to look good, too. I laminated the legs on my bench from three pieces of hardwood and left slots for each rectangular rung. And because I store my bench in the back of my van, along with my other tools, I attached the legs to the bench with butt hinges so that they fold flat beneath the bench. When I carry the bench up to my van, I set the front end in the back of the van then step out of the box. When I push the bench into the van, the front legs fold up almost automatically, and I fold the back ones up as the bench slides into the van.

Inside a door bench
A door bench typically holds an assortment of tools, but the depth of the bench is determined by the tallest tool. The sides of the bench have to be tall enough to clear the tallest tool so that a door will lie flat across the top of the box and not rock on a tool handle. As I said before, I step inside the box to carry it from one location to another, so the center of my box has a 20-in. clear opening. The remainder of the space in the box is filled by two 22-in. tool shelves. Nesting slots and holes are cut in each of the tool shelves for power tools and boring bits.

A door bench also requires an adjustable hook to hold the door in place while it's on edge. The hook has to be adjustable because doors vary in thickness from 1 in. for some screen doors and shutters up to 2½ in. or more for custom doors. The hook should be made from stout steel, like thick plate strapping. If the hook is mortised flush with the face of the box, the mortise acts like a track for the hook. After adjusting the hook for the thickness of the door, only one wing nut is necessary to tighten the hook down.

My door bench

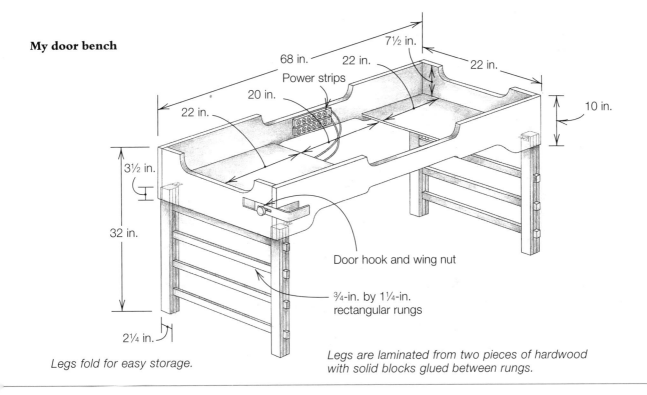

Legs fold for easy storage.

Legs are laminated from two pieces of hardwood with solid blocks glued between rungs.

HANGING DOORS

CUTTING THE TOP AND BOTTOM

After careful scribing, a door must next be carried to a door bench or a pair of sawhorses, so that the top and bottom can be cut. Carrying doors safely is also a matter of order, especially for heavy doors. As I describe the techniques I use for cutting and fitting a door, I'll also explain the strategy I use to move the door at my bench.

The easiest way for me to carry a door is on edge, over my shoulder, in the crook of my elbow, with my other hand holding the top edge steady. I always keep my back straight. When I reach my door bench I set the door down on its edge, spanning both ends of the bench. A pair of sawhorses can serve just as well as a door bench. Then I lay the door down flat. When I lay the door flat, I don't care which end is pointing toward the door hook on my bench, whether it's the top of the door or the bottom of the door. I only care about two things: First, I want the X up because my scribe lines are on that side of the door and I need to see them to cut the door; and second, I want the X toward me, on the same side of the box I stand on. After I finish cutting the top and bottom, it's easier to move the door into the next position if the X is toward me. You'll learn why in a minute.

Because circular saws cause tear out and chip end grain, before running the saw use a square and a utility knife to score a line across the face of the lock and hinge stile (see the left photo below). For flush doors with veneered skins, use a long straightedge and score a line completely across the top of the door. A Tru-Grip Clamp 'N Guide Tool is a handy straightedge, and it saves a few steps by eliminating the need for spring clamps (see the right photo below). A cutting guide (shown in the drawing on the facing page) works well, too, especially for beginners. Remember to also score the far edge of the door where the sawblade will exit.

To eliminate grain tearout, use a square on raised-panel doors and score a line across the grain before cutting the door with a circular saw.

Use a straightedge or Clamp 'N Guide to score a line across a flush door or a veneered door.

Cutting Guide

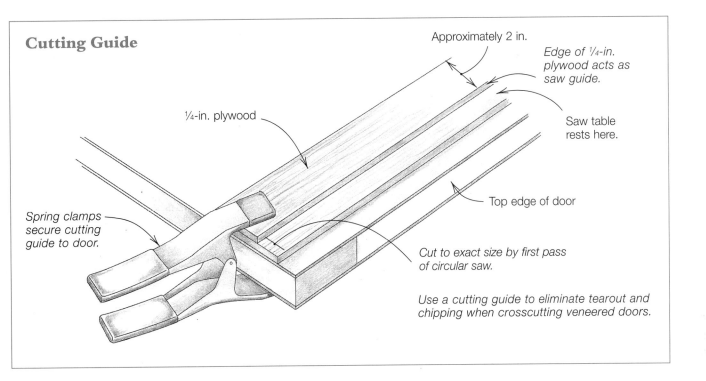

Approximately 2 in.

Edge of ¼-in. plywood acts as saw guide.

¼-in. plywood

Saw table rests here.

Spring clamps secure cutting guide to door.

Top edge of door

Cut to exact size by first pass of circular saw.

Use a cutting guide to eliminate tearout and chipping when crosscutting veneered doors.

The top of the door has to be cut very straight and the line followed exactly. If the saw tips, even slightly, it will be visible in the margin between the top of the door and the jamb. While anyone accustomed to a circular saw can cut the bottom of the door quickly, especially with the help of a guide, cutting the top isn't so easy. A good method is to cut just outside the pencil line, about ¹⁄₁₆ in., then to use a door plane to finish the cut right to the line. By holding the plane upside down you can see the pencil line and the cutter shaving the door (see the photo at right). Work in from one edge of the door and stop cutting before the cutter comes out the opposite edge. Running a plane out the end grain can blow a big chip off the stile. Leave that end of the door until last, then turn the plane right side up and make one or two passes to straighten the top of the stile.

With the door plane upside down, trim right to the scribe line. Stop before the cutter comes out the end of the door or it might tear out the grain. With the plane right side up, finish the cut by coming back in from that edge of the door.

HANGING DOORS

Single-Hinge Templates

Hinge templates and routers are a necessity for finish carpenters because economics won't permit the time to cut mortises with a hammer and chisel. It's difficult to make fixed multiple-hinge templates that are really accurate, but single templates are easy to make from scraps of wood or from damaged three-hinge templates.

To make a single-hinge template from scratch, use four pieces of wood. One wide piece forms the full-length back of the template. The template has to be long enough to support the router completely and still have room for a template nail at each end to secure the template to the door or jamb. The back of the template should be 2½ times longer than the router base, plus the length of the hardware. Most manufacturers make 4-in. and 3½-in. hinge templates with a 2¾-in.-wide opening, though only the height of the opening matters. Cut the two center pieces longer than necessary (you'll trim them off later, after the template is assembled). Lay out the height of the hinge near the center of the full-length back, then install the other two pieces 1/16 in. beyond each of those lines, leaving ⅛ in. for the additional width of the template guide.

Clamp the pieces together, then counterbore and fasten them with drywall screws. Don't glue them, since adjustments are often necessary. The fourth piece of wood is a narrow strip that determines the backset of the hinge. I often use steel in place of wood because it's stronger and won't deflect or crack and break. Most hinges are installed about ¼ in. from the door stop, so the last piece should be 3/16 in. thick. Counterbore for those screws, too, so that the screw heads won't rub on the door stop.

Ensure that the template will mortise the same backset on both the jamb and the door by installing a small stop on the bottom of the template. For 1¾-in. doors, the stop should be 1¾ in. from the outside edge of the template; for 1⅜-in. doors, the stop should be 1⅜ in. away from the outside edge of the template.

Most manufactured hinge templates are designed to work with a ⅝-in.-diameter template guide that fastens into the base of the router. The template guide collar surrounds the upper portion of the router bit and fits closely to the bit. The inside dimension of the ⅝-in. guide I use measures 17/32 in., which is only 1/32 in. larger than the ½-in. bit that spins inside. It's a close tolerance, but it works. I've never had a bit hit a template guide. Because the outside dimension of the template guide is ⅛ in. bigger than the router bit, the template is ⅛ in. larger than the hardware I'm installing.

Hinge templates

For use with ½-in. router bit and ⅝-in. template guide.

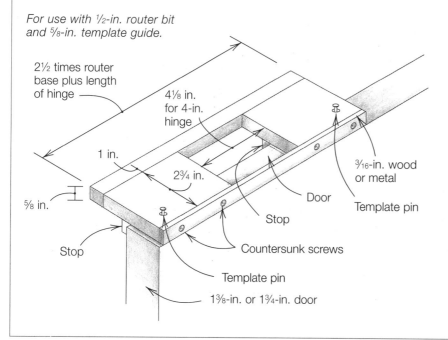

106 CHAPTER FIVE

PREPARING THE HINGE STILE

The next step is to move the door into a vertical position in the door bench and prepare the hinge stile.

The door must be on edge to plane the hinge stile and to install the hinges. Moving the door into the right position is easy because I made sure the X was toward me when I laid the door down on the bench. Now it's simple to tilt the door up, so that it's sitting on the edge of the lock stile. Reach inside the box and underneath the door to get one arm around the door, then lift it off the door bench. If the door isn't too heavy, hold it near the center and lower it straight down onto the rungs of the bench, with the X pointing toward the inside of the bench. If the door is too heavy, set one end down first, then slowly lower the other end onto the rungs. Slide the door forward until it's held securely by the door hook at the front of the bench.

The X and all the scribe lines must be pointed toward the inside of the door bench. That's one of the tricks to ensure that the hinges will be installed correctly. The door plane should always bevel in the same direction, down toward the inside of the bench. The only time I adjust my plane is to trim the top of a door, and because I do this so often, I scratched a line with my utility knife in the angle gauge on the plane. That way I can quickly change from a square cut to a 3 degree bevel.

Using a Door Plane

The Porter Cable 126 door plane is meant exclusively for doors and has the best depth-of-cut adjustment lever. Hold the handle and trigger of the plane in one hand. Use the thumb on your other hand to adjust the depth of cut, and use the other fingers to hold the fence tightly against the face of the door (see the photo below). If the fence isn't held against the face of the door, the bevel might vary and cause an ugly wave in the margin between the door and the jamb.

The Porter Cable 126 door plane is a workhorse. While the plane is running, a thumb lever smoothly adjusts the depth of cut.

Multiple-Hinge Templates

Multiple-hinge templates are manufactured to the length of a door and are meant to mortise two or more hinges in a single setup. These templates are available in fixed and adjustable types. Adjustable templates can be a major investment, so this short review should help you decide where to spend your money.

Most often I use a fixed multiple-hinge template made from dense ¾-in. plywood (far left in the top photo on the facing page). It's called a fixed template because the hinge locations are permanently cut into the template—perfect for new jambs and doors where it isn't necessary to match existing mortises. Jamb Fast and Templaco market similar templates in this design. I prefer the Jamb Fast (far right in the top photo on the facing page) templates because they're slightly thicker, stronger, and tend to last longer. When these plywood templates begin to age they sometimes crack and break. Before that happens and they're ruined, run the plywood template through a table saw and rip the thin inside edge off, then replace it with a piece of ¼-in. steel. With a steel edge, the template will never break.

Adjustable hinge templates are more expensive than fixed templates. I use an adjustable template whenever I have to match exisiting hinge mortises for more than two doors. Porter Cable manufactures an aluminum adjustable hinge template, as does Bosch. I prefer the Bosch design because it's simpler to set up (second from the left in the top photo on the facing page). Spaulding Industries manufactures a beautiful mahogany and aluminum template, called the VersaTemp (third from the left in the top photo on the facing page). Though I'm attracted to the look of this new product, it doesn't have the flexibility of the older all-aluminum models.

The VersaTemp has no adjustable hook at the top of the template, which is necessary for transferring the template from the jamb to the door and also helps to maintain proper clearance between the top of the door and the head of the jamb (see the bottom left photo on the facing page).

As with many tools, no single template has all the advantages. While the VersaTemp comes up short because it has no top hook, it does have superior hinge stops—the cylindrical bars that allow the template to adjust from 3-in. to 5½-in. hinges (in the bottom left photo, the pencil eraser is resting on the stop).

The hinge stops on the VersaTemp are secured by hex-bolt screws, whereas the stops in most other templates are held in place by pressure: The end of each stop is splayed like a small jaw and the jaw can be spread with a chisel or screw driver to tighten the stop in the template. Several times I've had a stop pop loose while routing a hinge mortise. Fortunately it's never happened on a stain-grade door because I tighten the stops before routing.

To change from a right-hand door to a left-hand door using a fixed multiple-hinge template, you just turn the template upside down. But an adjustable template is never inverted—the top of the template is always the top of the door. Bosch incorporates a swivel stop in its template that allows instant changes between right-hand and left-hand doors (point of the pencil on the bottom template in the bottom left photo). The VersaTemp design requires that two screws be removed and replaced each time the swing of the door is changed (point of the pencil on the top template), which consumes too much time for production door hanging.

Several companies, such as Bosch (second from the left) and VersaTemp (second from the right), manufacture adjustable hinge templates. Adjustable templates are much more expensive than fixed templates, like the one made by Jamb Fast (far right). I made the template on the far left from maple and attached a steel edge so it would never break.

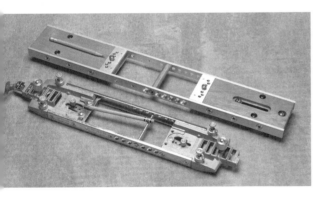

Adjustable hinge templates have to be adaptable to all thickness of doors, adjustable for all sizes of hinges, and work with both right-hand and left-hand doors. Some templates have adjusting pins that are easy to swivel, like the Bosch (bottom) and others have screws that have to be moved, like the VersaTemp (top).

Keep an eye on the scribe line. If the scribe line isn't parallel to the edge of the door, begin by slowly planing until the line and the edge of the door are parallel. As the plane moves along the door, adjust the cutter depth with your thumb. Once the edge of the door and the pencil line are parallel, plane closer toward the line with successive passes. Never bury the cutter or try to cut too much in one pass. Though the depth knob has settings from 1 to 3, I rarely dial down farther than setting 2. There's no hurry. Make smooth passes and slowly approach the scribe line. I like to just leave the line. That's how accurate a door planer will cut.

Finish with the plane by easing the edge of the door: Run the plane at an angle on the outside of the door then repeat the same technique on the inside (see the photo below). A plane with a new carbide cutter leaves a perfectly smooth surface, but after using the same blade to plane a few dozen doors, small lines, ridges, and even grooves will be left in the edge of the door. Rather than replacing the blade while it's still sharp, a quick pass with a belt sander will remove all imperfections.

Run the plane at an angle to ease the edges of the door.

HANGING DOORS **109**

Mortising for the Hinges

The most important thing to remember when laying out the hinge locations is to look for the X first. Because the scribe lines are always pointed toward the bench and the bevels are always cut down toward the bench, the top of the door might be at the top or it might be at the bottom of the bench. If you look for the X first, you won't hinge a door upside down.

Pull a tape measure down the edge of the door from the top to the bottom. If you recall, I suggested that as part of the scribing process you write the hinge measurements on the door near each hinge location. Now mark those measurements on the edge of the door with a sharp pencil and continue the lines across the edge of the door with a square.

Sometimes it's easiest to use a single-hinge template because the hinge layout probably won't match a manufactured multiple-hinge template. Manufactured single-hinge templates are inexpensive but they're also easy to make (see the sidebar on p. 106) from scraps of wood. Regardless of the template, the router has to be set deep enough for the thickness of the template and the hinge. Hinges are manufactured in different thicknesses (see Chapter 2), so I often have to check the depth of my router. One easy method is to hold one leaf of the hinge up against the back of the template and hold the router against the opposite side, with the bit extending through the template. Adjust the depth of the router until the bit is flush with the back of the hinge.

Position the template so it's $1/16$ in. above the first layout line—the template guide separates the router bit from the template by $1/16$ in. Remember to check the depth-of-cut setting before routing a complete hinge mortise. A shallow mortise is all that's needed to ensure that the hinge will be flush with the edge of the door. To avoid tearing out the face of the door, cut both ends of the mortise first by running the router bit into the door along each shoulder of the template. Then run a pass along the face of the mortise, and finally, clean out the center. For a smoother and more steady cut, move the router against the rotation of the bit, rather than with the rotation of the bit.

If I'm installing square corner hinges, I use a corner chisel to quickly square-up each mortise, and I use a Vix Bit to speed up drilling pilot holes (both of these items are available from woodworking suppliers). Always drill pilot holes for hinge screws. Driving four good-size screws into a space a little more than 1 in. wide and 3 in. to 4 in. long is like driving in a wedge! If pilot holes are not drilled, the stile will surely split and the door will be ruined.

PREPARING THE LOCK STILE

The hinge stile is finished once the hinges are screwed into the new mortises. Then it's time to roll the door and work on the lock stile. But it isn't necessary to lift the whole weight of the door straight out of the bench. Instead, grab the top edge just above the door hook and

Lift the end of the door up out of the hook. Let the back of the door rock down to the ground on the rear rung.

Checking for the Right Bevel

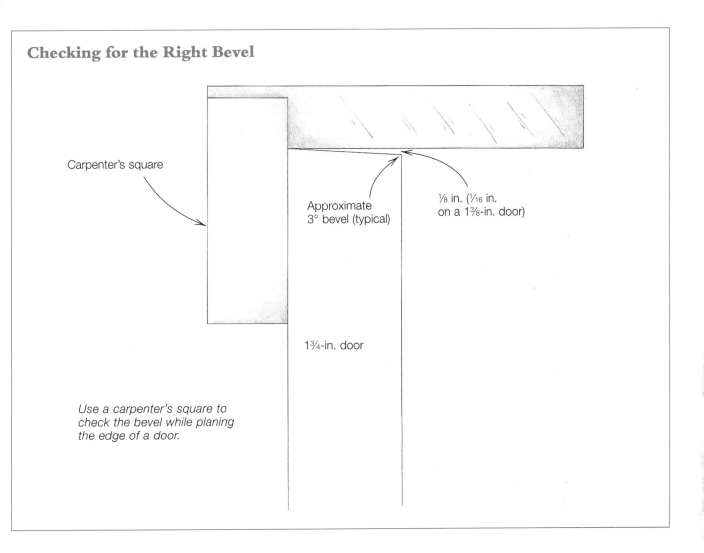

Use a carpenter's square to check the bevel while planing the edge of a door.

lift the door up and out of the door hook, as shown in the photo on the facing page. Let the back end of the door pivot on the rung of the bench until it touches the ground. Continue lifting while the door goes past vertical and down into your other hand. Now, for just a moment, it's necessary to carry the full weight of the door—but only for a few steps. Set the door back down near the front leg of the door bench. Then set the back end down onto the back rung of the bench.

Planing the lock stile involves the same steps as planing the hinge stile. If you're new to using a power plane, try practicing on a clean 2x4 before you even use it on a door. Learning to use a door plane, like learning to ride a bike, can't be taught with words—it has to be done. Planing a consistent bevel is one of the biggest hurdles most beginners face. If the bevel is inconsistent the door will never fit the opening properly: The door will rub against the jamb in some places and may not even shut, or the gaps will be too big in other places. Check the bevel frequently, too. While planing the door, place a square across the bevel at various points along the lock stile. On a 1¾-in. door with a bevel of about 3 degrees, the bevel should fall away from the square about ⅛ in. (see the drawing above). A 1⅜-in. door should fall away from the square 1/16 in. from the high side to the low side of the bevel.

As soon as the edge is planed, beveled, and sanded smooth, measure down from the top of the door for the lock location (see the photo below). The X is now down near the rungs, but still look for it every time you lay out a lock. The last thing you want to do is drill the door upside down—a mistake that can't be fixed, even in a paint-grade door. I've never seen anyone successfully fill a 2⅛-in. hole in a door—a year later there's always a faint circle cracked in the paint.

Boring for a Lockset

Standard locksets, also called bored locksets, require a 2⅛-in. hole through the face of the door and a horizontal 1-in. hole through the edge of the door. Residential locks usually have a 2⅜-in. backset—the distance from the leading edge of the door to the center of the face bore. A 2¾-in. backset, common in many commercial applications, can't be used on a 4-in. door stile—the trim will overhang the back of the lock stile, and on a French door you risk drilling into the glass.

The method used to bore for locksets is determined by the tools you have. Relatively inexpensive lock-boring kits are available that include a hole saw and a spade bit. The hole saw should be a multitooth cutter, not a single-blade or adjustable cutter. For installing an occasional dead bolt or a few locksets, a hole saw is fine, but for drilling lots of doors, a hole saw is too slow and a lock-boring jig is a necessity. Everyone begins with a hole saw, though, and I still use one on special occa-

Always look for the X (not visible in this photo). Measure down from the top of the door and mark the lockset location.

sions, such as for boring metal doors, for special backsets, and for dummy locks where an astragal often interferes with a boring jig.

Careful layout is always the first step when using a hole saw and a spade bit to drill a door. The paper template that's included with every lock works well and guarantees an accurate layout. A square works, too, but just be sure to measure carefully. The photos on p. 114 illustrate the step-by-step process. My best advice is to drill the 2⅛-in. face bore from both sides of the door to avoid tearing out the grain. Once the face bore is drilled, the spade bit only has to penetrate 1¼ in. into the door, and the risk of the bit accidentally wandering out the face of the door is minimal.

Installing Locksets

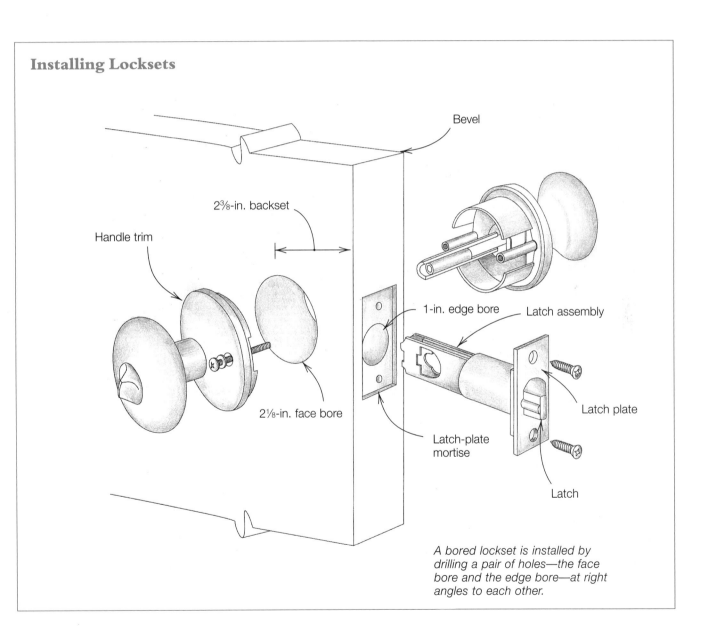

A bored lockset is installed by drilling a pair of holes—the face bore and the edge bore—at right angles to each other.

HANGING DOORS 113

Use the paper template that comes with the lock and a sharp awl or pencil to locate the center of the hole. The paper template will also locate the 1-in. latch hole centered in the edge of the stile. (Photo by Rich Ziegner.)

Always drill the 2⅛-in. face bore hole first. It's easy to start a hole saw level and straight—just be sure the teeth on the saw are all cutting evenly when you begin, then hold the drill steady and go slow. (Photo by Rich Ziegner.)

Don't push the hole saw too hard, and don't run the bit too fast. If you feel the teeth cutting, you're going at the right speed. Stop when the mandrel bit (tip of the pilot hole) penetrates the opposite side of the door. (Photo by Rich Ziegner.)

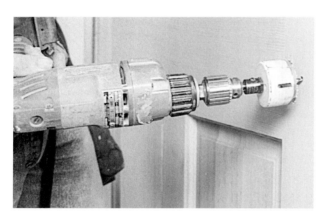 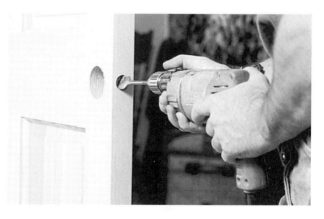

Finish the hole from the opposite side. Use the hole to guide the bit into the door. Keep the hole saw running at the same rpm, but gently guide it into the door so that it cuts smoothly and doesn't jerk out of your hand when it joins the first hole. (Photo by Rich Ziegner.)

Hold the drill motor level and square to the door before starting the spade bit. Use a sharp spade bit and run it fast but push lightly and evenly; apply just enough pressure so that the bit cuts straight into the door. (Photo by Rich Ziegner.)

Drill the face bore first, so the edge bore is easier.

Using a Boring Jig

With a boring jig, there's no risk of drilling out the face of the door. The only concern is locating the right spot to drill. To find the location of the lockset on the door, measure down from the top of the jamb to the center of the strike location and subtract 1/8 in. for the margin between the top of the door and the head of the jamb (see p. 101 for more information). For flush doors, that measurement would be the center of the lockset on the door, and could now be transferred to the freshly beveled edge of the door. But for stile-and-rail doors, the lockset should center on the lock rail. As I said earlier, I'd rather move the strike on the jamb and fill the old mortise than install the lockset off center on the door.

A boring jig is easy to use once the center of the lockset is marked on the edge of the door. Place the 1-in. edge bore bit on the center of the mark and lower the jig onto the door. Tighten the large clamp snugly. (An adjustable bolt applies just the right amount of pressure on the back of the jig so that the jaws on the front close evenly across the lock stile.) Then double-check the backset. Two spring-loaded backset stops are easily adjusted from 2 3/8 in. to 2 3/4 in. Double-check that they're set properly. Remember, a 2 3/4-in. backset will not fit in a 4-in. stile.

Drill the face bore first, as shown in the photo above, then pull the spur bit out of the jig before beginning the edge bore. Too many spur bits have been ruined by

HANGING DOORS 115

Lock Boring Made Easy

A boring jig eliminates having to lay out the locations for the edge bore and the face bore, and because each bit spins in a large bushing, there's no concern about keeping the bit or the drill motor straight. A boring jig uses a 2⅛-in. spur bit for drilling face bores. A spur bit has a short, triangular tip at its center, small cutters around its circumference, and one cutting blade. These bits cut holes fast when they're sharp, but steel bits get dull quickly. Carbide-tipped spur bits are far superior, but much more costly. It doesn't pay to buy carbide unless you drill a lot of doors. Nordic Saw and Tool currently charges more than $100 to retip the standard 2⅛-in. steel spur bit that's included with most boring kits or almost $200 for a new carbide-tipped spur bit.

The 1-in. edge-bore bit supplied with most boring jigs has a replaceable spade-bit tip. The spade bit works very well if it's kept sharp; it's easy to sharpen the bit on a belt sander or with a file. Alternative 1-in. spur bits are available, but they tend to clog and never cut as fast as the original 1-in. spade bits.

Boring jigs also come with a quick-release driver that snaps over the hexagonal end of each bit. Though it's called a quick release, it can't be released quickly; the bit has to stop spinning before it's safe to touch the driver. If you're like me and don't like waiting, use a standard ⅜-in. socket extention. It's simpler and safer and actually releases quickly.

Because Schlage and Baldwin dead bolts require a 1½-in. face bore, a 1½-in. multispur bit also comes as standard equipment with all boring jigs. A 1½-in. reducer ring comes with the jig, too, and has to be inserted before using the 1½-in. bit. The reducer ring prevents the back of the door from chipping or tearing out. To bore doors quickly, I use a ½-in. high-speed drill and push hard when I'm drilling. Sometimes the bit cuts through the back of the door suddenly, but I've never damaged a door using a boring jig. This is one tool I recommend highly.

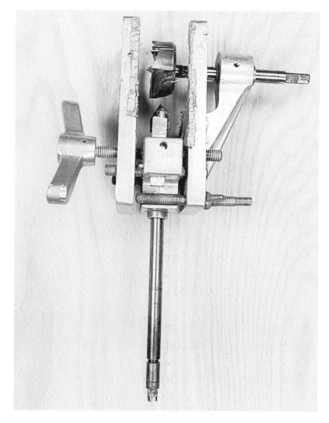

A boring jig is like a portable double drill press and many companies manufacture them, including Templaco, Jamb Fast, and Classic Engineering, shown. (Photo by Rich Ziegner.)

accidental bit collisions. Drill the edge bore next. After the edge-bore bit clears the face-bore hole, continue drilling for another ½ in. Many locks, like Baldwin and Schlage dead bolts, have long latches and require deep latch holes. When drilling for a deep latch in a French door or a stile-and-rail door without a lock rail, drill that extra ½ in. slowly. Watch the bit carefully (see the photo at right). If it penetrates the back of the stile on a French door, it will break the glass.

Mortising for the Latch

Lockset and dead bolt latches have to be flush with the edge of the door. The mortises are quick and easy to cut with a sharp chisel (see the photos on p. 118). To avoid splitting the door or chipping out more wood than you want, always cut the top and bottom of the mortise first, then work on the sides and center. And never drive a chisel hard into the direction of the grain, especially alongside the edge of a door. The grain near the edge of a door splits easily, so just tap the chisel lightly.

I prefer to use a router and homemade template for all latch and strike mortising because a router is much faster and a template ensures that each mortise will be the perfect size and depth. I use the same router for these templates as I do for my hinge templates. I make my latch templates thicker so that the depth of the router bit doesn't have to be changed from the hinge setting.

Most residential latches are 1 in. wide and 2¼ in. tall, so only one template is necessary (Schlage does make a dead bolt with a wider 1⅛-in. latch face, though it's rarely used in residential work). A router template is easy to position: Simply place it on the door over the latch bore. Stops on the bottom of the template keep it centered over the edge of the door (see the sidebar on pp. 120-121), but the template still has to be centered over the latch bore. Manufactured templates are also available, and some, like Templaco and Jamb Fast, come with centering devices, but I scored a line across the center of my template, which serves the same purpose. After routing the mortise, use a corner chisel to square up the corners (see the photo on p. 119).

For longer latches and dead bolts, drill the edge bore ½ in. deeper than the face bore. But don't drill too deep. If there is no lock rail, the bit will come out the back of the stile.

HANGING DOORS   117

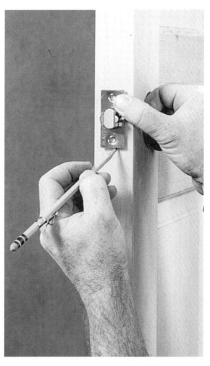 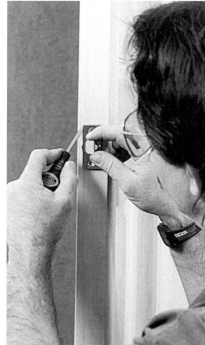

Far left: Insert the latch in the edge bore and use a sharp pencil to trace the outline of the latch. (Photo by Rich Ziegner.)

Left: Use an awl to score with the grain—along the sides of the latch—and a utility knife to score across the grain. Score the inside of the mortise with a chisel. Hold the blade of the chisel at a 45 degree angle from the door and square across the grain. Tap the chisel into the mortise every ¼ in. to avoid tearout. (Photo by Rich Ziegner.)

Far left: With the bevel of the chisel toward the center of the mortise, drive the chisel just inside the top and bottom lines. The bevel will push the chisel right to the line. With the bevel of the chisel against the door, chisel out the mortise from the bottom up toward the center, then from the top down toward the center. (Photo by Rich Ziegner.)

Left: Finish the mortise by laying the bevel of the blade flat near the center of the mortise and tapping the chisel gently toward the shoulder cut. Check the fit and shave a little more if necessary. (Photo by Rich Ziegner.)

A corner chisel speeds up squaring the mortise.

HANGING THE DOOR

Now that all the machining is complete, you're ready to carry the door back to the jamb. While there's nothing wrong with boring a door after it's been installed, I prefer to bore most doors at my door bench. Working at my bench is much faster because all the tools are there and plugged in, and I don't have to drag everything to each opening, along with a power cord. I also don't have to clean up a mess of sawdust at every opening, and that saves time, too. This practice is especially productive when hanging new doors in existing jambs, with carpeted floors and furniture everywhere and where jambs are already mortised for hinges and strikes.

Swinging and Adjusting the Door

There are several ways to hang a heavy door alone. Because pilot holes have to be drilled in the jamb for all the hinges screws, one method is to pull the top hinge pin and split the top hinge. Use the loose leaf to locate the pilot holes in the jamb for the two bottom hinges, then attach the leaf to the top mortise. Lift the door onto that leaf and drive in the hinge pin. The second method is to tilt the door in the opening, with the bottom of the door farther into the opening than the top, until one of the holes in the top hinge lines up with the corresponding hole in the jamb (see the photo on p. 122). Once one screw is driven through the hinge, push your foot against the bottom edge and screw off the remaining hinges. If a door is scribed carefully and

HANGING DOORS **119**

Making Router Templates

I make my router templates from scraps of hardwood. I use the same method to make lock templates as I do for hinge templates—instead of cutting a hole in a large piece of wood, I get a crisp, straight template using strips of wood (see the drawing on the facing page). Almost all latch plates are 1 in. wide and 2¼ in. tall, so the template I use most often has an opening that's 1⅛ in. wide and 2⅜ in. tall.

It only takes four pieces of wood to make the template. The two outside pieces can be almost any size—mine are about 1¼ in. by 14 in., but the two short inside pieces have to be ripped to the width of the latch plate, plus ⅛ in.

Use screws to assemble all the pieces. It's rarely necessary to make adjustments, but these templates can also be used to cut mortises for odd pieces of hardware. Screws allow each template to be changed to a custom size quickly, using the base and template pins already installed.

Lay out the template opening near the center of an outside piece. Clamp the two inside pieces to the layout lines, then drill countersunk pilot holes and

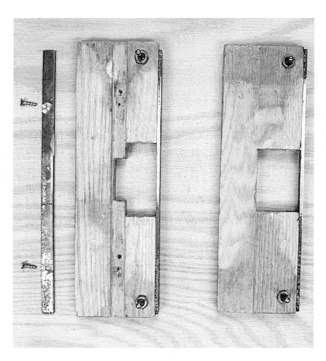

Strike-plate templates: On the right a template for Westlock and Kiwkset type "Lip" strikes. On the left a template for Schlage and Baldwin "T" strikes with a removable metal strip (far left) that allows the template to fit both lockset and dead bolt strikes. (Photo by Rich Ziegner.)

Top of standard latch-plate template, and bottom, with guides attached. (Photo by Rich Ziegner.)

drive in the screws. The other flanking outside piece should be clamped, too, so that the pieces remain flush before driving in screws. Cut the ends square and install template pins in the center of the top and bottom. Position the template pins far enough from the opening so that the base of a router won't bump into them.

I attach alignment stops to the back of all my latch templates, too. The stops are positioned to fit snugly around a 1¾-in. door and to center the template on the edge of the door. I don't hang that many 1⅜-in. doors, but when I do, I trace the outline of the latch on the edge of the door, then center the template by eye.

Making a standard latch-plate template

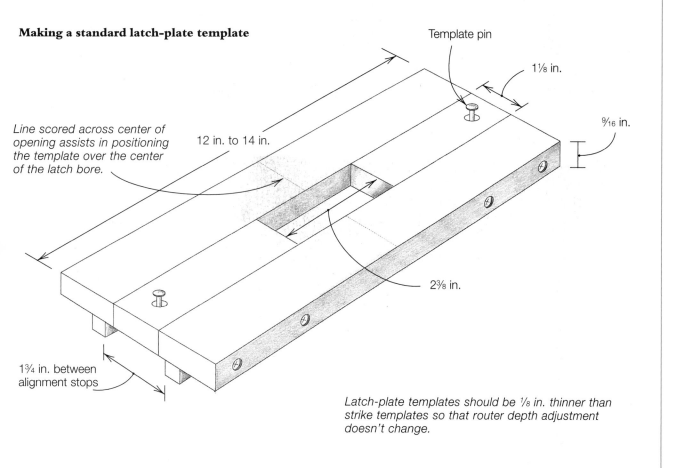

Latch-plate templates should be ⅛ in. thinner than strike templates so that router depth adjustment doesn't change.

If the door is too heavy to lift, tip it in the opening and drive one screw through the top hinge, then use a foot to push the door plumb and screw off the remaining hinges.

beveled correctly, it will usually fit the first time up. If it doesn't fit just right, small hinge adjustments will probably correct the problem. But always adjust the hinges before installing the lock strike because the position of the strike might change.

The secret to adjusting hinges is to imagine the door and the jamb as a rectangle within a rectangle. Sometimes one rectangle needs to be rotated a little—and that's always the door—so that both rectangles are square to each other. This adjustment is common and necessary when the head gap is a little tight above the lock stile. A door might even rub a little on the head jamb, directly above the lock bore (see the drawing on the facing page). When this occurs, the gap between the jamb and the lock stile, near the top of the door, is often too big. The quickest remedy is to bend the top hinge and rotate the door down. Place the head of a nail set between the leaves of the hinge, and pull the door closed on the nail set, squeezing the hinge gently. Don't pull on the door too hard or the hinge screws will rip out of the jamb. Try the fit. (I usually don't have time to shim paint-grade hinges, but if the hinges have a special finish, I take the time to cut and place small pieces of cardboard behind each hinge, rather than bending them, which often leaves dents.)

Conversely, when the head gap on the strike side is too big, the space between the jamb and the lock stile at the bottom of the door is often too big. If this is the case, place the nail set in the bottom hinge, and gently squeeze the door shut. Then try the fit.

If these two situations are reversed and the lock stile of the door is rubbing on the jamb, then the hinge gap is often too big. Now the hinges have to be squeezed shut a little. Squeezing the hinges pulls the door in the opposite direction, back toward the hinge side of the jamb. Drive the hinge pin up until it engages only the top hinge knuckle. Tighten a crescent wrench on the top knuckle of the hinge that's attached to the door—not the leaf attached to the jamb. Pull and bend the top

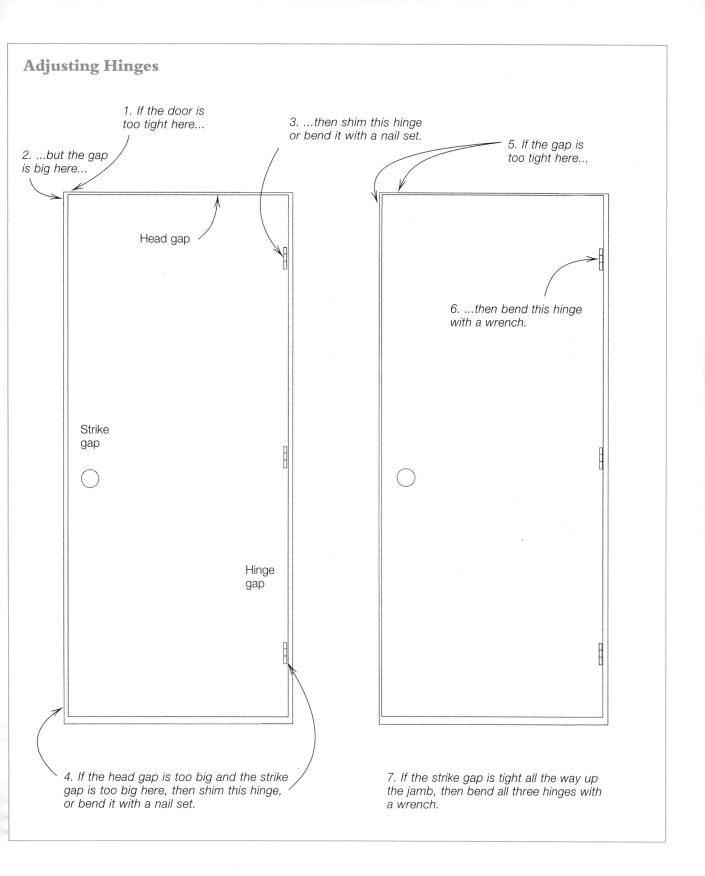

Drive the hinge pin up until it engages only the top knuckle. Tighten the crescent wrench snug on the knuckles attached to the door—not the leaf attached to the jamb. Pull the wrench in the direction of the lock, and bend the knuckles just a little.

Insert the locator, point out. Close the door and push the locator into the jamb. Open the door and immediately circle the dimple made by the locator, so that the strike plate isn't installed over an old nail hole.

knuckle on the door toward the lock stile (see the photo at left). Bend each knuckle attached to the door just a little. This adjustment will pull the door tighter to the jamb, and increase the strike gap (see no. 5 and no. 6 in the drawing on p. 123).

Locating the Lock Strike

There are two ways to locate the right position for a lock strike. The first way is quicker, but not always exact. Most boring jigs come with a device called a center marker—a steel cylinder with a sharp point centered on its face (see the bottom photo at left). I also call it a locator because 90% of the time it accurately locates the center of the lock. To use this device, slide it into the edge bore and close the door. Push the point of the locator hard into the jamb. If it doesn't feel like it's marking the jamb sufficiently, then use a pair of pliers or a pry bar for leverage.

The locator marks a dimple in the jamb. Find and circle the dimple with a pencil right away. The locator ensures that the strike hole and the latch will align properly; the pencil mark guarantees that you won't mistake some other mark or nail hole for the dimple and drill a 1-in. strike hole in the wrong place.

While a locator is an accurate tool most of the time, the second method I recommend for locating lock strikes always works, which is good because some strikes are not adjustable. Most strikes have a small strip of metal, bent at 90 degrees, which projects out the back of the strike and into the latch hole. The metal strip acts like a finger and can be bent to pull the latch toward the door stop. Some Baldwin and Schlage strikes don't have that metal finger and those strikes can't be adjusted to pull the door against the stop when the latch engages the strike. Non-adjustable strikes have to be installed perfectly the first time, and this method always works. Place a piece of masking tape over the face of the strike—pencil lines show up clearly on masking tape—then place the strike on the edge of the door, on top of the latch. The latch shouldn't touch the shoulder of the strike but should just clear the opening in the strike. With a pencil riding the edge of the door, trace a line down the face of the strike (in the left photo on the facing page, note the adjustable metal finger on the back of the adjustable strike

Hold the strike over the latch, so that the latch just clears the hole in the strike. Let the pencil ride along the edge of the door and trace a line across the strike.

Close the door. Ride a pencil along the edge of the door and trace a line on the jamb. Also mark the horizontal centerline of the latch.

I use to illustrate this procedure). Shut the door and trace a line on the jamb with the pencil riding the edge of the door the same way (see the top photo at right). Pencil a horizontal line across the jamb, centered on the latch, and hold the strike centered on that horizontal line. Line up the vertical pencil marks on the strike and the jamb, then trace the outline of the strike on the jamb (see the bottom photo at right).

A spade bit is the safest thing to drill into an old jamb, just in case you run into a nail. For new wood, a Forstner bit cuts a cleaner hole with a flat bottom, which looks better. Don't drill completely through the jamb for a lockset strike unless the hardware includes a dust bucket (the molded plastic or brass insert that provides a finished background within the strike hole). Once the strike hole is drilled, mortise for the strike plate.

Rather than a chisel, I prefer to use a router and template for this job, because it's cleaner, faster, and eliminates having to hammer on the jamb, which could change the margin between the jamb and the door. I attach a strike-plate template to the jamb and rout out the mortise following the same procedures used when mortising the door. The last step is to install the strike and close the door. The dull thunk of the latch falling into the strike, just as the door comes up against the stop, is sweet.

Place the strike on the jamb, with the vertical pencil marks in line and centered on the horizontal line. Trace the outline of the strike.

Chapter 6
ADVANCED DOOR HANGING

HANGING A PAIR OF DOORS

INSTALLING MORTISE LOCKS

Some of the material in the previous chapters may have been a review for experienced carpenters because I included basic instruction so that readers would understand the fundamental skills of hanging doors. Many of those skills are essential to understanding the advanced techniques in this chapter, beginning with hanging a pair of doors and installing flush bolts, extension flush bolts, and mortise locks. Furthermore, many of the techniques in this chapter will be required to understand procedures in the following chapter on specialty doors.

Much of what I've learned about doors has been gleaned from production door hangers in southern California, who butt, bevel, and bore 20 to 40 doors a day. There are surely skilled craftsmen who may scoff and disagree, who believe that a door should be fit slowly, carefully, and painstakingly. But the end result—the perfect fit of a door—is the primary consideration.

HANGING A PAIR OF DOORS
Scribing and cutting a single door so that it fits the first time up is a good challenge for a carpenter, and it's a challenge easily achieved through the methods I described in the previous chapter on hanging a single door. But to me, hanging a pair of doors so they both fit the first time in the opening is a real challenge. The door hangers I learned from, the hangers on my crew, and many other door hangers I've met in the field pull this trick off every day, and the only secret is careful scribing.

Scribing a Pair of Doors

Setting up to scribe a pair of doors takes a little more time than setting up to scribe a single door, but the result is worth it. Begin by clipping both doors to the jamb with door hooks, then start shimming each door gradually. If the jamb has an oak sill, then place a few blocks of wood under the doors equal to the height of the sill. Small shims can be used for final adjustment. Line up the heads of the doors to the head of the jamb. Time can be saved if the tops of the doors don't have to be cut, which means it's worth the effort to shim the doors until the margin between the tops of the doors and the head of the jamb is even, straight, and slightly more than $1/16$ in. Often, existing jambs on remodels are not straight and the heads of both doors have to be cut, no matter how much care is taken during the scribing setup. If the heads of the doors are flush, then the panels or lights will probably be in a straight line too, but don't worry about that yet.

Once the heads are flush, adjust the sides of the doors. With the lock stiles touching in the center, move the doors laterally, as if they're both one big door, until an equal amount can be scribed off each hinge stile. Use a small prybar to move the doors in fractional increments. After the doors are perfectly centered, check that the lock rails are aligned and level (see the right photo below). I use a transit or builder's level whenever I install a series of French doors to ensure that all the muntin bars (the molding around the glass panes) will be in a straight, level line.

Hanging a pair of doors is a challenging job, but the process is really the same as hanging a single door. Always begin with careful scribing.

The measurement from the jamb head to the center lock rail on each door should be the same. Adjust the shims beneath the doors until the muntin bars, lock rails, or molding line up perfectly.

ADVANCED DOOR HANGING

Overlapping Astragals

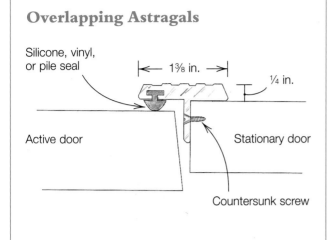

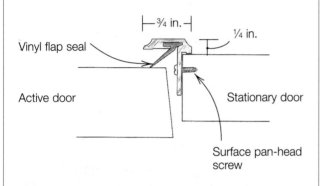

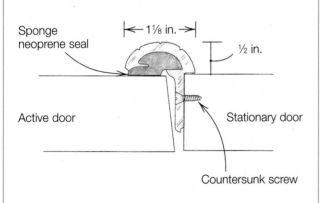

Astragals are available from most weatherstripping manufacturers in anodized bronze, aluminum, gold, bright-dipped gold, and white. Some are also available in milled bronze/brass and stainless steel.

Some jambs are not set plumb, so choosing the right position for the doors can be frustrating. Usually it's best to split the difference between hanging the doors perfectly plumb and level and hanging the doors perfectly parallel to the jamb. If a pair of doors is installed perfectly plumb in an out-of-plumb jamb, then the hinge stiles and the heads of the doors will be cut at an angle. If a jamb is severely out of plumb, those angled cuts will be obvious. The angles can be hidden somewhat if a compromise is made between perfection and reality.

Marking active and stationary doors I use a simple X to locate the hinge stile on a single door, but a pair of doors has to be marked differently. One door must be marked ACT, because that door will receive an active lockset and probably a dead bolt; the other door must be marked STAT, because it will have an astragal and some type of bolts to hold it stationary, either flush bolts or surface slide bolts. Opinions vary on which door should be active: Some believe that the right-hand door, looking from the outside, should be the active door; some believe that the right-hand door, looking from the inside, should be the active door. I recommend looking for the electrical switches. Light switches should never be behind the active door, because a person entering would have to open the door and walk all the way around the door to switch on a light.

Scribing for astragals and thresholds Having marked which door is active and which is stationary, the next step before scribing the doors is setting up for the astragal (see the drawing at left). For overlapping aluminum astragals that are installed after the doors are finished (along with the remainder of the weatherstripping), leave the two doors touching in the center, with the lock stiles tight against each other, then scribe all four sides of the opening—the heads, stiles, and bottoms. But if the astragal has to be installed now, such as a wooden T astragal or an aluminum astragal with factory-installed flush bolts (see Chapter 2), then spread the lock stiles apart the thickness of the astragal so that no material is planed from the leading edge of either door. Lock stiles are often only 4 in. wide, and most locks will take up nearly all of that depth, which brings up the confusion about astragals.

The nature of astragals can be ambiguous and opinions are generally split along two sides. One side believes that the thickness of the astragal should be removed from the lock stile of the stationary door because the thickness of the astragal should be counted as part of the stationary door. The other side believes, as I do, that an astragal is an isolated molding, forming a distinct and separate line between a pair of doors. The point is proven by the location of the dummy hardware in the face of the stationary door. The dummy hardware should have a backset equal to the active hardware, but if the stationary door is planed ¾ in., then the dummy hardware will either be too close to the astragal or too close to the back edge of the lock stile. Of course, if the jamb has been built wider to accommodate the thickness of the astragal, then very little wood will have to be removed from either door (see Chapter 2), and the argument is unnecessary.

To scribe for an astragal, slide each door over just a little, so that the doors are spread apart the thickness of the astragal (see the photos below). Because astragals measure anywhere from ½ in. to ¾ in., and sometimes even more, measure the astragal carefully so that you remove just the right amount from the hinge stile of each door. Once the doors are centered and separated for the astragal, scribe the head and sides as you do for a single door. Hold the scribes perpendicular to the jamb, and spread them slightly more than ⅛ in. apart for the hinge stiles; squeeze the scribes closed for the heads.

While the doors are hooked inside the opening, it's a good idea to scribe the bottoms, too. An exterior door requires some type of threshold and door shoe (see Chapter 8). By knowing the type of threshold and door shoe, you can scribe the doors at the proper height now rather than having to take them down again later. Thresholds and door shoes, like hinges and locksets, should be on the job site when the doors are installed. Once the doors have been scribed on all four sides, I move them to my bench. I like to tackle the stationary door first, because it's the hardest.

After spreading the doors apart with a small prybar, use a tape measure to check that the gap between the doors is even from the top to the bottom.

Move each door an equal distance, then measure from the jamb leg to the hinge stiles and check that both doors are still centered in the opening.

ADVANCED DOOR HANGING

Installing Astragals and Flush Bolts

I spend more time processing the stationary door than I do working on the active door because fitting the astragal correctly and installing flush bolts takes time. When I started out as a carpenter in northern Arizona, astragals and flush bolts were things I tried to avoid, like washing my truck. Most doors were prefit, and flush bolts were someone else's problem. But when I moved to southern California, I ran into more pairs of entry doors and more pairs of French doors, and nothing was prefit. I had no choice but to learn about astragals and flush bolts.

The face of the astragal should project past the bottom of the door and reach the top of the threshold. Cut the leg of the astragal flush with the bottom of the door so it won't prevent the door shoe and vinyl sweep from being flush with the front edge of the door.

Installing an astragal at a door bench is very simple as long as you follow these two rules: Always make sure the hinges are pointing toward you, and always make sure the face of the astragal is pointing away from you.

Before installing an astragal it's essential to know what type of threshold and door bottom are being used, otherwise the astragal cannot be cut for length, and the position of the flush bolts can't be determined exactly. If the doors are getting standard ½-in. door shoes, cut the astragal ½ in. longer than the length of the door, so that the astragal will reach down past the doors to the threshold and thoroughly seal the opening. The astragal should be a permanent part of the stationary doors, so glue and nail the astragal to the door, but don't drive any nails near the lockset or the dead bolt locations, and stay clear of the flush bolt locations, too.

Installing a beveled wooden astragal Some wooden astragals, like the standard lumberyard variety (see the photo on p. 36) are beveled. I've seen beveled astragals installed every which way: Sometimes the beveled edge is attached to the stationary door, sometimes it's facing the active door. Some installers even believe that using a beveled astragal saves having to bevel the active door. But it doesn't. If the astragal is installed at an angle, or with the bevel facing the active door, the shoulder of the astragal will push the active door proud of the stationary door (see the top drawing on the facing page). The best way to install a beveled astragal is to bevel the stationary door before installing the astragal. Place the astragal on the edge of the door and use a small square to check that the face of the astragal is square to the active door.

Installing a Simple-T astragal There's no easier way to install flush bolts, especially extension flush bolts, than by using a Simple-T astragal, because it comes with built-in ¼-in.-diameter, 12-in.-long, solid-brass slide bolts. Any carbide-tipped circular saw blade can cut easily through the entire astragal, brass and all. Installation is extremely fast and simple. Measure the distance between the top of the door and the top of the threshold, subtract ⅛ in. for clearance, and cut the astragal to that length. Always cut everything off the bottom of the astragal so that the top flush bolt is as long as possible and easy to reach. And always check that the flush bolt is retracted before cutting the astragal to fit.

Installing a Beveled Astragal

Incorrect installation

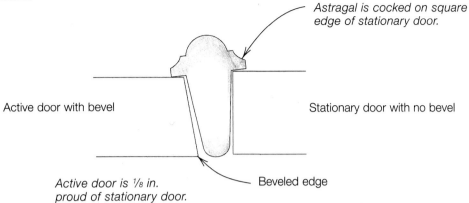

Astragal is cocked on square edge of stationary door.

Active door with bevel

Stationary door with no bevel

Beveled edge

Active door is 1/8 in. proud of stationary door.

Correct installation

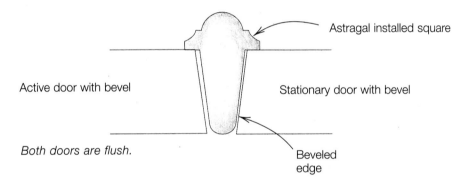

Astragal installed square

Active door with bevel

Stationary door with bevel

Both doors are flush.

Beveled edge

Beveled astragal and square astragals

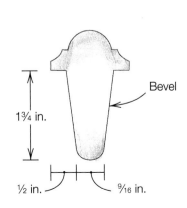

Bevel

1 3/4 in.

1/2 in.   9/16 in.

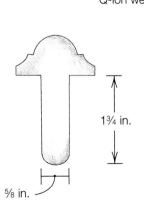

1 3/4 in.

5/8 in.

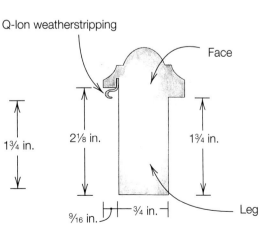

Q-lon weatherstripping

Face

2 1/8 in.

1 3/4 in.

9/16 in.   3/4 in.

Leg

ADVANCED DOOR HANGING   131

Anodized Aluminum Locking Astragals

Mortise astragal

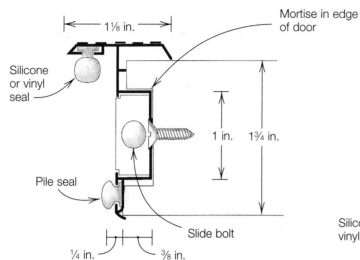

All three aluminum locking astragals have mechanical slide bolts. Spring slide bolts are available for each model. Spring slide bolts are activated by a pressure-sensitive touch switch located on the slide bolt linkage, within easy reach even on tall doors.

Oak fascia cap for slimline astragals

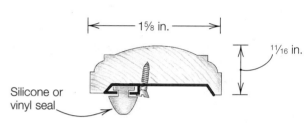

Colonial surface-mounted astragal with thermal break

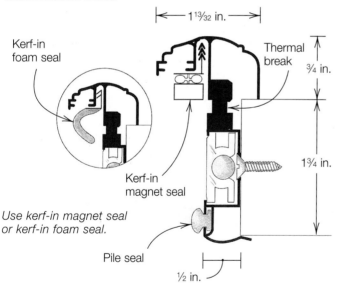

Slimline surface-mounted astragal

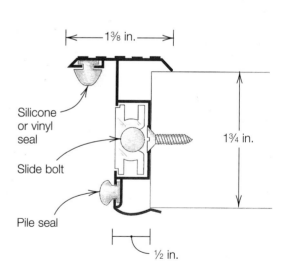

Use kerf-in magnet seal or kerf-in foam seal.

A thermal break astragal is essential for cold climates. Otherwise, frost would penetrate and collect on the inside of the astragal.

Simple-T astragals can also be installed after the doors are swinging, without having to remove the stationary door from its hinges. That means the gap between the stationary door and the active door can be checked with both doors up, so after final planing the astragal can be secured to the stationary door. The same can be done with a regular wooden astragal, but the stationary door has to be taken down to install the flush bolts, which is why I install astragals and flush bolts before hanging a stationary door on its hinges.

Installing an aluminum locking astragal There are two basic choices in aluminum locking astragals: surface-mounted astragals that require no special preparation and mortise astragals that require a full-length dado in the edge of the inactive door (see the bottom left drawing on the facing page). Before I owned a plunge router I used a regular router equipped with a standard flush-bolt fence to cut this dado. I dialed the bit down after each pass. Now I use a plunge router. The fence on my router base is set to center a ¾-in. bit in a 1¾-in. door, leaving approximately ½ in. of wood on both sides of the flush-bolt mortise. To cut a 1-in.-wide mortise, I attach a ⅛-in. strip of scrap to the face of the fence, which draws the router bit closer to the face of the door and leaves ⅜ in. of wood. I run my router up one edge of the door, making several passes to reach a depth of ⅜ in. Then I place the fence on the other side of the door and run the router again. This ensures that the mortise is perfectly centered in the edge of the door.

The dado can also be laid out with a set of scribes and cut with a circular saw. On a 1¾-in. door, spread the scribes far enough to drag a line on the edge of the door, leaving ⅜ in. of wood on each side of the edge. The space between the lines should equal 1 in. Set the circular saw blade to cut ⅜ in. deep. Start the mortise by cutting right up to each pencil line. Clean out the center of the mortise after the outline is cut precisely. I've cut many mortises this way but it's not enjoyable. Cutting dadoes and mortises with a circular saw is messy and time-consuming, the cut is never clean, like a router cut, and often the dado has to be shaved and adjusted with a chisel.

Once the dado is cut for a mortise-type astragal, installing the astragal is the same as installing a surface-mounted model. Aluminum astragals invariably have to be cut to length. If the weatherstripping includes standard ½-in. door shoes (Chapter 8), then the length of the astragal should be cut equal to the height of the door. The ½-in. gap between the bottom of the astragal and the top of the threshold will be sealed by a separate neoprene sponge bottom seal included with the aluminum astragal (see the drawing on p. 134). After the astragal is installed, the bottom seal is glued into the bottom of the astragal.

If an interlock threshold and hook strip are installed, then the overall length of the astragal, including the bottom neoprene seal, should be the distance from the top of the threshold to the top of the door. Regardless of how much needs to be cut from the length of the astragal, always cut excess material off the bottom. The top of the astragal should never be cut except for the notch that's necessary to allow the astragal to clear the jamb rabbet.

Measure the length of the astragal, then remove the snap-in covers and the sliding bolts before cutting it. The covers and the bolts have to be removed to secure the astragal to the door. On models with spring bolts, four set screws must be loosened to remove the spring bolts. Aluminum locking astragals cannot be installed while the door is swinging, like a Simple-T, because the spring bolts slide in from the top and bottom, which is why I install astragals with flush bolts before hanging the stationary door on the hinges.

I cut aluminum astragals on my chop saw with a carbide blade, although a hacksaw will do the job, too. Once the astragal is cut to length, fasten it to the stationary door with screws. Replace the sliding locks and insert the neoprene sponge seal in the bottom of the astragal with a dab of adhesive, then hang the door back up on the hinges.

Installing 6-in. flush bolts I use three dependable methods to install common 6-in. flush bolts. I'm sure there are more ways, but these techniques have worked well for my crew and me. This first method is quick and doesn't rely on a template or jig. Instead it depends on

Installing Aluminum Locking Astragals

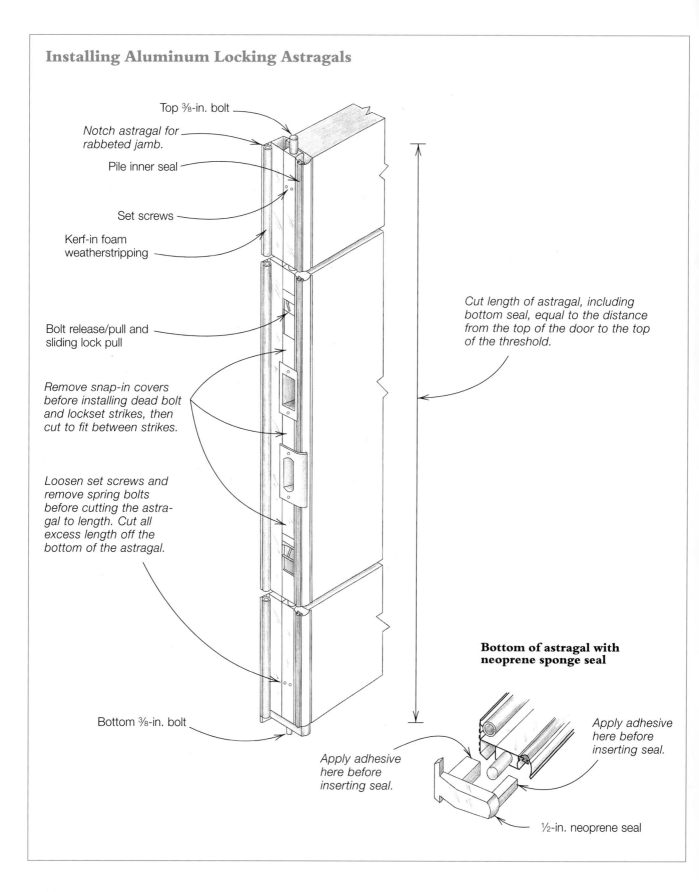

a steady hand and a circular saw. Trace the outline of each flush bolt on the edge of the door, as shown in the photo at right, and draw a mark across the outline below the top screw, at the top of the retracted bolt linkage. Set the blade on a circular saw to cut 7/8 in. deep, then carefully cut just inside the outline, stopping at the cross mark below the top screw. Once the side cuts have been made right to the lines, make repetitive cuts to remove the wood that remains in the center. Finish cleaning the mortise with a chisel.

Making fine cuts with a circular saw is difficult, especially when the blade blocks the line, which always happens on the far side of the mortise. For that reason, the second method I use depends on a router. Manufactured templates are available for flush-bolt installation, but they're expensive, they're never the right size or configuration, and the thickness of most of them prevents the router bit from cutting the full depth of the mortise—a drill or circular saw always has to finish the job. Instead, I devote one router to flush bolts. If having a router just

Trace the footprint of the flush bolt and draw a line across the outline, just beneath the top screw, where the bolt linkage ends. From that line up, the mortise is only 1/8 in. deep for the face of the flush bolt; but from that line down it's 7/8 in. deep for the linkage and the bolt.

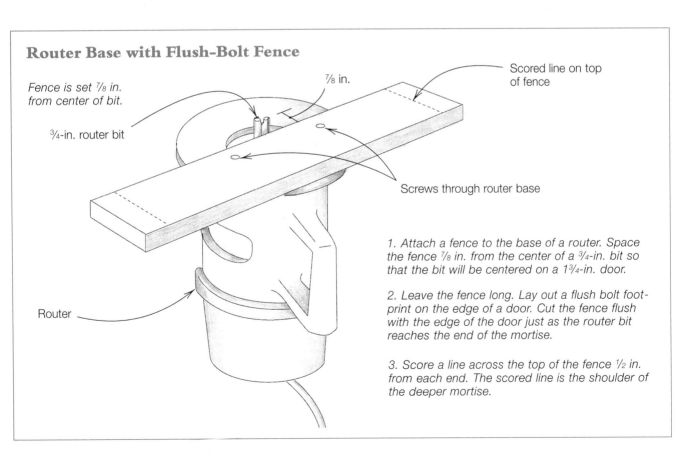

Router Base with Flush-Bolt Fence

1. Attach a fence to the base of a router. Space the fence 7/8 in. from the center of a 3/4-in. bit so that the bit will be centered on a 1 3/4-in. door.

2. Leave the fence long. Lay out a flush bolt footprint on the edge of a door. Cut the fence flush with the edge of the door just as the router bit reaches the end of the mortise.

3. Score a line across the top of the fence 1/2 in. from each end. The scored line is the shoulder of the deeper mortise.

ADVANCED DOOR HANGING

for flush bolts doesn't make any sense (unless you hang a lot of doors), then buy an extra router base. I have several bases for my routers so that I don't have to change elaborate router setups or fences. The fence for a flush-bolt setup centers a ¾-in. router bit in a 1¾-in. door. No template guide is necessary, and flush bolts don't have to be traced on the astragal. The fence rides the edge of the door and after each pass, the depth of the bit is increased fractionally until the mortise is finished.

Some time ago I grew tired of dialing my router down for every pass, so the third method, and the one I use most often, is a plunge router (see the photos on this page). I use the same fence setup as with the fixed router, though I installed threaded inserts in the fence so I can quickly attach or detach it with two machine screws. The longer bits in my plunge router also solved the problem I had installing flush bolts in wooden astragals. Because the router has to ride on the top of the astragal, standard routers can't reach the bottom of the cut, but a plunge router can cut almost 2 in. deep.

Installing extension flush bolts Installing extension flush bolts can be quick and easy, too, but the trick is to use templates. Making good templates isn't difficult and doesn't require much time—an hour or two at most. And with a little care, templates can be used over and over, year after year, and last almost indefinitely.

A flush bolt fence can be attached to a regular router, but a plunge router works the best. The scored line that's flush with the edge of the astragal marks the end of the deepest part of the mortise.

End grain is difficult to chisel, especially so close to the edge of the astragal. Instead, hold a router on its side to cut the mortise in the end of the stile.

For wooden astragals, I attach a removable spacer to the side of the fence so that the router rides level (on top of the astragal) and the bit is perpendicular to the door.

Two-Step Templates

Two-step templates are a must for mortises that have two different levels—a shallow mortise in which the trim plate seats and a deeper pocket for the linkage. Two-step templates are also useful for large lock strikes that require deep dust buckets, and they're wonderful for Soss hinges (see Chapter 7), pocket door pulls, and electrical jamb switches.

I build these templates the same way I build all my templates—by cutting separate pieces for the width of the mortise, then joining the pieces so that they're spread apart the length of the mortise. The inside dimension of each template must also include the offset between the cutting edge of the router bit and the outside of the template guide.

For my extension flush bolt template, I used a piece of ¼-in. steel for the outside strip, because it had to be thin yet strong. I used two 3¼-in.-long screws as stops for the router bit guide. The screws thread through the hardwood rail from the back side of the template, so they're always accessible. The location of each screw is determined by the required step in the mortise.

Extension flush-bolt template

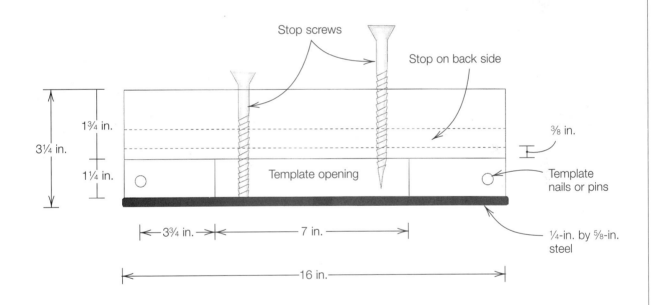

Template opening is ¼ in. larger than the flush bolt for a ¾-in. bit and a 1-in. template guide in a plunge router. For standard routers with a ½-in. bit and ⅝-in. template guide, the template opening should measure 1⅛ in. by 6⅞ in.

ADVANCED DOOR HANGING

With the extension flush bolt retracted, mark the door near the back of the bolt linkage to determine the end of the dado that must be cut in the door.

Through the use of a template, a difficult, time-consuming job not only becomes foolproof and fast, but perfectly accurate every time.

I use two methods for installing extension flush bolts: one method for doors that have wooden T astragals where the flush bolts are installed in the leg of the astragal, and a different method for doors that have overlapping aluminum astragals where flush bolts are installed directly in the edge of a door. I use the same extension flush bolt template for both methods, but I prepare the deep hole for the extension bolt much differently.

The bolt hole required for an extension flush bolt is centered ¾ in. below the top edge of the door, which also happens to be where the astragal and the edge of the door are joined. This is not an inviting spot to bore a 10-in.-deep hole because the drill bit tends to wander in

Scribe a line down each side of the stile, marking a ¾-in.-wide mortise in the center of the door.

Set a router or circular saw to cut to the depth of the extension bolt, but subtract the thickness of the astragal. Clean out the remainder of the mortise with a sharp chisel.

a joint between two pieces of wood. Instead of drilling a hole for the extension bolt and before attaching the astragal to the door, plow a dado in the stationary door with a router or circular saw, as shown in the photos on the facing page.

After the dado is cut, attach the astragal with glue and nails, but don't drive any nails into the area of the dado. Next, lay out the location of the extension flush bolt, being certain that the flush bolt is retracted (if the bolt is extended the rod won't be long enough to engage the jamb). The pocket mortise for an extension flush bolt can be cut with drill bits and chisels—even a circular saw can be used for the shoulder cuts, but a router and two-step template (see the sidebar on p. 137) are quicker and more dependable, as shown in the photos on this page.

After tracing the footprint of the extension flush bolt trim plate, attach the template centered over the outline of the trim plate. The stop screws should be retracted for the first pass of the router, which is only the depth of the trim plate, about 3/16 in.

Drive the stop screws into the template before mortising the deeper pocket. A plunge router works best for this operation, though a regular router will do the job, too.

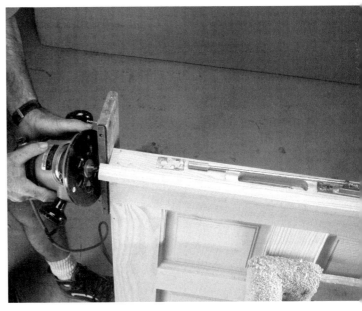

Use a standard dead bolt strike template to mortise for the retaining plates that have to be installed in the top and bottom of the door. A router works best for cutting this mortise because it's all end grain and very close to the edge of the astragal.

ADVANCED DOOR HANGING

Drill a 2-in.-deep hole with a ¾-in. spade bit first, then attach a lock-boring jig to the edge of the door.

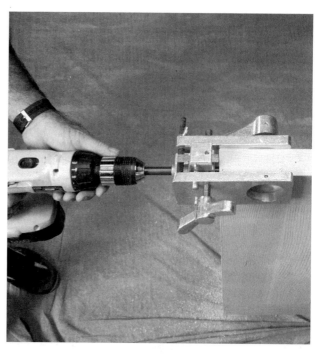

A ½-in.-diameter drill bit fits perfectly in the boring jig bushing. Use a 12-in.-long drill bit and bore into the door to the depth of the bit.

Remove the boring jig and finish the hole freehand. Because only the first 2 in. of the bit are fluted, the previously bored hole will act as a guide and the bit won't wander.

Installing extension flush bolts directly into the edge of a door is an entirely different matter. A 10-in.- to 12-in.-deep hole must be drilled because the door can't be plowed. Boring deep holes in doors can be frightening, but a lock-boring jig removes the risk (the process is shown in the photos on this page). Using even longer bits, I can drill all the way through a door for electric strikes that require a hole from the lock strike to the hinge.

There are several ways to install the flush bolt into the pocket mortise. The hardest way is to set the linkage in the pocket mortise and then to try to find it with the threaded extension rod. An easier way is to thread the extension rod into the back of the linkage, then to insert the rod and the linkage in the pocket mortise, without the barrel of the flush bolt attached (see the photo on the facing page). Thread the barrel onto the rod after the linkage and the rod are installed. Retract the flush bolt and adjust the barrel so that it's flush with the top of the door and doesn't fall behind the retaining plate, then attach the retaining plate. Be sure to drill pilot holes for every screw—splitting a door at this late stage can be a terrible loss.

Completing a Pair of Doors

As I said earlier, it takes time to install astragals and flush bolts, which is one reason why hanging a pair of doors is a lot more work than hanging two individual doors. Pairs are also more work because bigger openings mean bigger problems, especially if the jamb isn't set perfectly. But before adjusting the doors to fit the opening, notch the top of the astragal to fit under the rabbeted jamb head.

Shut the stationary door and scribe a line across the face of the astragal, just below the rabbet in the jamb head. If the jamb is kerfed for foam weatherstripping, then the line should be 1/8 in. below the bottom of the jamb head. If rigid weatherstripping is being applied to the surface of the jamb, then the scribe line on the face of the astragal should be 5/16 in. below the jamb head (see Chapter 8). Use a small backsaw or dovetail saw to cut along the line, but stop cutting just before the blade touches the door. Finish the cut from above, flush to the face of the door.

The neoprene seal on the bottom of a locking aluminum astragal should be in contact with the threshold. Because the inactive door of a pair is rarely opened, it doesn't matter if the seal is tight against the threshold, as long as there's no light penetration.

Now both doors can be closed and the gaps adjusted. As with a single door, shim or bend the hinges until the gaps between the doors and the jamb, and the strike gap between the two doors, are equal and straight. If, after adjusting the hinges, the active door is rubbing on the stationary door, then plane a little off the lock stile of the active door. It's safe to plane up to 1/8 in. on standard 4-in. lock stiles. But if the active door is projecting past the stationary door by more than 1/8 in., then the hinge stile will have to be planed because some locksets are almost 3 5/8 in. deep.

Thread the rod halfway into the linkage, then insert the rod in the pocket mortise. Bend the rod slightly so that the back of the linkage doesn't scrape the door just before it drops into the mortise.

After adjusting the hinge, head, and strike gaps, check the cross-leg before locating and drilling for the flush-bolt strikes. Wide openings exaggerate cross-leg. If the walls or jamb can't be moved to bring the legs of the jamb parallel, then try reducing the cross-leg by moving a few hinges. If the bottom of the active door isn't shutting completely but the top of the active door is, then move the top hinge on the active door away from the stop. Loosen the screws on the center hinge first, then remove the screws from the top hinge. Let the weight of the door pull the top hinge out of its mortise just a little, then drill new pilot holes with a Vix bit or fill the holes with stick matches and reattach the top hinge. Sometimes the top hinge only has to move 1/8 in. away from the stop in order for the two doors will lie flat against each other. If the situation is reversed and the bottom of the active door is touching the astragal but the top isn't, then move the bottom hinge away from the stop.

ADVANCED DOOR HANGING **141**

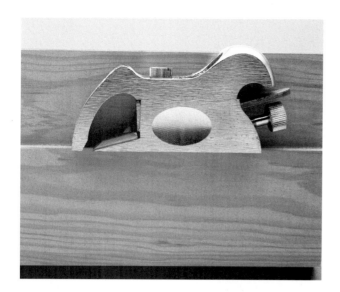

If the cross-leg is severe, the rabbeted stop on one jamb may have to be planed. Use a bullnose plane that fits tight against the shoulder of the rabbet and can cut all the way from the sill to the jamb head. Make slow, even passes.

Avoid moving any single hinge more than ⅛ in. because it will be noticeable—the gap between the stop and the door will be difficult to seal, and the door will be proud of the jamb near the hinge that's moved.

In some cases correcting cross-leg requires more work than moving hinges. If a top or bottom hinge has been moved ⅛ in. away from the rabbeted door stop and the doors still don't meet flush at the astragal, then the opposite hinge may have to be moved *closer* to the stop. Moving a door closer to the stop necessitates planing the rabbeted stop on the jamb. I use a bullnose plane (Stanley Bullnose No. 90) for this job, as shown in the photo above. First draw a straight line on the face of the rabbeted stop. The line should cover the shoulder of the rabbet ⅛ in. at the end of the door that's not touching the astragal and taper to zero at the end of the door that is touching the astragal. Usually the line must taper in at the head of the jamb, too, so draw a tapered line across this area using the same dimensions, and plane the rabbeted stops on the head and the leg flush at the top corner. Planing such a small amount off the rabbeted stop isn't difficult with a sharp plane, and it's never noticeable.

After planing the rabbet, remortise one hinge closer to the stop, and move one hinge away from the stop, thus dividing the distance of the cross-leg and bringing the two doors flush to each other. The hinge that's been moved away from the stop will no longer cover its hinge mortise, so protect that hinge with masking tape and fill the exposed hinge mortise with wood filler. Minwax Wood Filler is a good choice for stain-grade jambs because it takes stain well and has adhesive qualities that auto-body filler lacks.

Locating flush-bolt holes and strikes The location of the flush-bolt holes and strikes can also be adjusted to correct a little cross-leg, though only a small amount of lateral pressure should be applied to a flush bolt (too much pressure can cause a door to split); besides, moving a flush-bolt hole more than ⅛ in. is noticeable—the doors will no longer be parallel and straight to the threshold. I use a homemade locator (see the sidebar on the facing page) to pinpoint the exact position of flush-bolt holes, and I use a spade bit to drill the holes. While drilling the holes, I wobble the bit back and forth slightly, parallel to the jamb head, and create an oblong hole that matches the hole in the strike. The bolt is sure to fall in the hole every time, and the door is certain to be drawn tight to the jamb.

The mortises for the flush-bolt strikes are easy to chisel by hand, but I use a template and a router. I try to avoid driving a chisel into a jamb because the jamb might move and crack the paint or ruin the gap between the doors and the jamb, especially if shims haven't been installed near the mortise. I recommend tracing the outline of the strike over the flush-bolt hole and positioning a template directly over that outline. Drive the template pins into the jamb carefully, and clean out the round corners of the mortise carefully, too, so that the jamb doesn't move. And always drill pilot holes for the small mounting screws.

A Foolproof Flush-Bolt Locator

To use the locator, first extend the top flush bolt just enough so that it enters the hole in the locator, but not so far that it interferes with the door clearing the jamb. Shut the door against the door stop and hold the locator still.

Retract the flush bolt and open the door, but don't let the locator move. Use a sharp pencil to trace the ½-in. outline of the flush-bolt hole.

I made my flush-bolt locator from a thin strip of sheet metal. I cut the metal strip 1½ in. wide and about 2½ in. long, then drilled a ½-in. hole close to one end and a ¼-in. hole close to the opposite end. I use the ½-in. hole for standard flush bolts and the ¼-in. hole for Simple-T astragals. I bent the strip of metal in the middle, just a little, so that my fingers can hold the locator and still clear a swinging door.

Before locating the hole for the bottom flush bolt, lock the top flush bolt and check for cross-leg again. Position the bottom bolt so that the two doors are flush, even though it may mean pulling or pushing the stationary door just a little.

ADVANCED DOOR HANGING **143**

Installing Baldwin Mortise Locksets

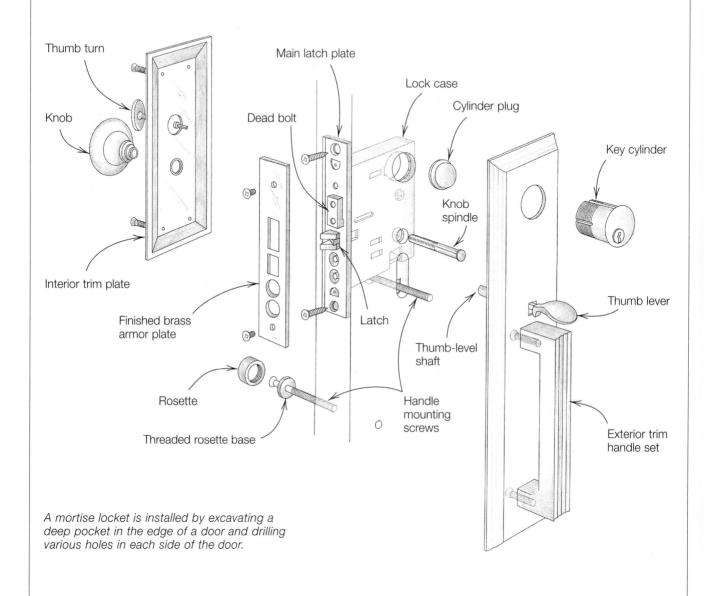

A mortise locket is installed by excavating a deep pocket in the edge of a door and drilling various holes in each side of the door.

INSTALLING MORTISE LOCKSETS

Mortise locks exemplify the relationship between carpentry and magic. Besides requiring a deep pocket mortise, these locks necessitate holes of various sizes on both sides of the door for knobs, handles, levers, and a dead bolt. The trick to laying out and drilling a door for a mortise lock is having the right tools and techniques—and of course practice—but the secret is templates. In the following section I'll discuss the simple tools, techniques, and templates that I use.

Before setting up any tools, check the lock and all the parts. Be certain that the lock is the correct hand and that all the parts are included, before lugging any tools to the opening. Open the hardware boxes and hold the lock case up to the door to check that the bevel on the latch is pointing in the right direction. Baldwin locks are not reversible and if the lock case is the wrong hand, it can't be installed. Other mortise locks, like Bouvet and Schlage, are easy to reverse on the job site (more on that later).

While holding the lock up near the door, check the backset, too. Baldwin lock cases are manufactured in 2½-in. and 2¾-in. backsets, which is the distance from the leading edge of the door to the center of the lever or knob and dead bolt. A 2½-in. backset case is 3½ in. deep and requires a pocket mortise that's 3⅝ in. deep. But a case for a lock with a 2¾-in. backset is 4 in. deep and requires a pocket mortise that's 4⅛ in. deep. In other words, a 2¾-in.-backset lock case cannot be installed in most stile-and-rail doors because the lock stiles are only 4 in. to 4⅛ in. wide. Schlage locks aren't commonly found on residential jobs because all Schlage L locks are manufactured with 2¾-in. backsets and the cases are 4⅜ in. deep. While checking the backset of the lock case, also measure the lock stile on the door and check that the door hasn't been planed too much. Some carpenters hang doors by hinging them first, then they finish fitting the door by planing everything from the lock stile. But a 2½-in. backset Baldwin lock cannot be installed if the lock stile has been planed down to less than 3¾ in.

After ensuring that the lock case is correct, check the following parts:

- **Key cylinder.** The key cylinder should be long enough to reach through the face of the exterior trim plate to the center of the door. Baldwin manufactures cylinders in several lengths because trim designs vary in thickness. Extended cylinders are available for custom doors that are more than 1¾ in. thick. If the cylinder is too short, it won't reach the linkage inside the lock case; if the cylinder is too long, it will hit the opposite side of the lock case and stick when the key is turned.

- **Threaded mounting screws.** Two threaded mounting screws are necessary to secure exterior handle sets. These screws have to be long enough to penetrate the door and reach the threads on the exterior trim. Doors that are thicker than 1¾ in. require longer screws. A polished rosette and threaded rosette base should be included for the bottom mounting screw. (After checking the two mounting screws, measure the distance on the exterior handle set from the top mounting screw hole, just beneath the thumb lever, to the bottom mounting screw hole in the bottom of the handle, and save that measurement for later use.)

- **Knob spindle.** Be sure the knob spindle has been included because on newer locks, once the case is installed in the pocket mortise, the spindle has to be inserted through the door and the case before the exterior trim and dead bolt cylinder can be installed.

Checking these parts is also important because it confirms the necessary drilling pattern for the door. Some mortise locks are patio styles and have only a lever or knob outside and no exterior handle set. And some patio locks have no exterior key cylinder. So before laying out any holes, check the hardware carefully.

While all the parts are spread out, take care of a few essential steps. Packaged along with the dust bucket is the cylinder plug, a large brass threaded washer, about the size of a half-dollar. This must be threaded into the interior side of the lock case, opposite the key cylinder. The cylinder plug equalizes the pressure on the cylinder

ADVANCED DOOR HANGING

After centering the trim in the lock rail of the door, mark the top and bottom of the lock case. The marks should be 6¼ in. apart. (Photo by Rich Ziegner.)

yoke and pins. Rotate the plug until it's flush with the box and the notch in the plug lines up with the yoke pins, then apply a piece of tape to secure the plug so that it won't jiggle out of position.

Before starting any installation steps, I also like to cut the exterior thumb-lever shaft and the interior thumb-turn shaft to length to fit the thickness of the door and the depth of the exterior trim, so that when I begin assembling the lock trim, I'll be able to move along quickly and need fewer tools in my hands. To determine how much must be trimmed from the back of these two shafts, first measure across the front edge of the door, near the pocket mortise, and make a small mark ⅛ in. beyond the center of the door. On a 1¾-in. door, that mark will be 1 in. from the face of the door. Next, place the exterior handle setup against the face of the door and hold the thumb lever down so that the shaft is level. Use a utility knife to score a line across the thumb-lever shaft at the pencil mark and cut the shaft with a hacksaw at a slight angle—about 3 to 5 degrees—down and away from the face of the door. That slight angle will prevent the end of the shaft from rubbing on the shoulder of the pocket mortise inside the door. The interior thumbturn shaft can be trimmed square by following the same steps.

The Pocket Mortise

Locksets are normally located between 34 in. and 38 in. high off the floor and centered in the lock rail of the door. Mortise locks are no different, although sometimes it is difficult to decide which part of the lock should be at the most comfortable level—the exterior thumb lever or the interior knob. Therefore, begin laying out the pocket mortise by holding the exterior handle set against the door, with the thumb lever about 36 in. from the floor. Raise or lower the exterior trim plate slightly, so that it centers on the lock rail of the door, then mark the location of the thumb-lever shaft. Next, hold the lock case against the side of the door and mark the position of the interior knob, then check that the interior trim plate is acceptably centered, too. On custom doors with unusual lock rails, compromise is the key. After locating the position of the lock case, draw a line across the edge of the door, at the top and bottom of the case, to locate the position of the pocket mortise (see the photo at left).

Mortising the door is always the first step in installing the lockset, and there are several ways to cut the deep pocket for the lock case. Regardless of which method is used, the case should slip in and out of the door without snagging. Any pressure on the lock case can interfere with the function of the lock.

One method for cutting pocket mortises is using a spade bit and a set of sharp chisels. I've used this method many times in the past, though it's risky and not recommended for the faint-hearted. Baldwin lock cases are 7/8 in. wide, so a 1-in. spade bit is preferable for this job. Trace a line down the center of the door, between the two marks made previously for the top and bottom of the lock box. To help keep the spade-bit tip right on the line, use an awl and score the line deeper. For extra caution, measure and lay out each overlapping hole on the edge of the door, then drive an awl into the center of each hole location. Also, trace a line 3/8 in. in from each edge of the door, so that it's quickly apparent if the spade bit begins to wander. Finally before drilling, place a piece of tape on the shaft of the bit as a stop guide so that all the holes will be drilled to a uniform depth and the bit won't penetrate the back of the lock stile.

After all these precautions are taken, hold the drill motor firmly with two hands, square to the door, and drill each hole slowly, beginning with the top and bottom of the mortise (see the left photo below), then drill the overlapping holes. The more the holes overlap, the less chiseling will be required (see the right photo below), but the holes can't overlap too much or the bit will wander into a preceding hole. I've known carpenters who also clamp blocks of wood to both sides of the door, to support the thin stock. This is a good precaution, too. There are no bad precautions; in fact here's an even better one.

As you may have noticed, I use my lock-boring jig for many drilling jobs besides installing bored locksets. Before I bought a lock mortiser, I often used my boring jig for mortise locks because it eliminates the risk of drilling with a spade bit. Also, because the boring jig drills straighter holes and uses a 15/16-in. bit, the walls of the mortise aren't quite as close to the face of the door.

When cutting pocket mortises, drill the top and bottom holes first, then drill the overlapping holes. Hold the bit firmly to the center line, so the shoulders of the mortise will be straight. (Photo by Rich Ziegner.)

Use a wide sharp chisel to pare the shoulders of the mortise. Shave the mortise clean, especially near the very bottom of the pocket. (Photo by Rich Ziegner.)

Using a lock-boring jig allows the holes to be closely overlapped, which reduces chiseling. When chiseling, be sure to clean the bottom of the mortise too, so that the lock case will slip in and out without binding.

The method I now prefer to use for cutting mortises is a lock mortiser. These machines cost about $800, but like many tools, if they're used frequently they pay for themselves quickly. I install numerous mortise locks every year, sometimes several in a single home, which makes the investment worthwhile. My Rockwell mortiser (now manufactured by Porter Cable) has a 1½-hp router cradled between four tubular guides. Two of the guides are parallel to the door and allow the router to travel up and down; the other two guides are perpendicular to the door and control the depth of cut. A carbide cutter head is attached to the business end of the extended spiral router shaft. Turning a large crank drives the linkage that raises and lowers the router and also engages a gear that moves the router closer to the door with each complete pass. The cut of the router is carefully controlled, about ¼ in. with each pass. An adjustable bolt acts like a stop so that the router won't cut too deep.

Though they look intimidating, lock mortisers are fairly easy to use. Clamp the mortiser to the door, aligning the cutter head with the top mark of the lock case on the edge of the door. Crank the handle and lower the cutter toward the bottom mark. Adjust the position of the

When using a lock mortiser, push the button on the gear box before turning the crank handle so that the mortiser will automatically control the depth of the cut. (Photo by Rich Ziegner.)

Slowly turn the crank clockwise. Keep one finger on the off switch, just in case you need to turn it off quickly. (Photo by Rich Ziegner.)

mortiser on the door until the cutter head is aligned with both marks. The mortiser's clamps automatically center the cutter head in the edge of the door. Start the router motor, engage the gear box, and turn the crank slowly (see the photos on the facing page). These machines throw out handfuls of sawdust (always directed right at your face), and they make an awful roar, so wear ear and eye protection when using one.

Mortising for the latch plate Once the pocket is cut and cleaned out, slide the lock case in and trace the outline of the main latch plate onto the edge of the door. Even though I use a template and router to cut the mortise, tracing the outline of the latch plate, with the lock case in the mortise, ensures that everything will align perfectly. The main latch plate mortise should be ¼ in. deep, which also allows room for the finished armor plate.

Besides mortising the face of the door for the latch plate, the shoulders of the pocket mortise have to be notched for the latch trigger pin. Even though the latch trigger is only pinned to the outside of the lock case (on 2¾-in. backset locks, the trigger is on both sides of the lock case—see the drawing on p. 155), both shoulders of the pocket mortise must be notched or the lock case won't seat completely (see the photo at right). This often-overlooked detail will result in the latch hanging up inside the lock case and the lock failing to function smoothly, if at all.

As you notch for the latch spring pins, be certain that the notch on the outer shoulder completely clears the latch spring pin in the lock case, or the lock won't work. (Photo by Rich Ziegner.)

Laying Out the Cross-Bore Holes

The next step is to lay out the locations for the interior knobs and levers, the exterior key cylinder and thumb lever, and the mounting screws. Years ago I made a quick and simple plywood template (shown in the drawing on p. 150) to locate these cross-bore holes, which saves time and worry, because all these different-sized holes can get confusing, and a mistake can be costly. Today, Baldwin markets a beautiful lightweight aluminum template that wraps around the door and guarantees accuracy (shown in the photo on p. 151). The Baldwin template is available from most hardware stores and sells for about $20. I use the aluminum template most often, though there are times when I still rely on my old plywood template.

Always position a plywood template flush with the exterior face of the door and ⅛ in. proud of the interior face (because of the bevel on the edge of the door). Be certain the top-of-mortise marks are aligned exactly with the top of the latch plate mortise, then secure the template with a spring clamp. The U-shaped Baldwin template wraps around the door and has enough spring to hug the door tightly. Two metal fingers within the

A Plywood Layout Template for Cross-Bore Holes

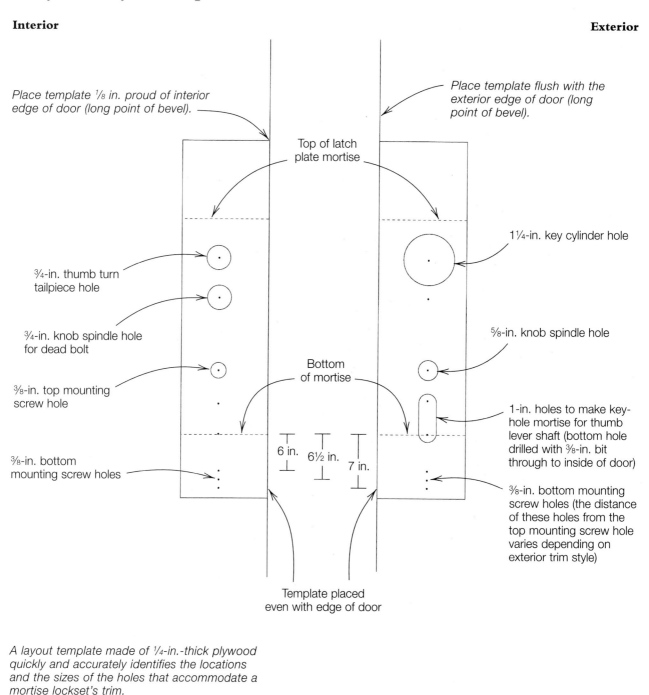

Interior — Place template 1/8 in. proud of interior edge of door (long point of bevel).

Exterior — Place template flush with the exterior edge of door (long point of bevel).

Top of latch plate mortise

1¼-in. key cylinder hole

¾-in. thumb turn tailpiece hole

¾-in. knob spindle hole for dead bolt

⅝-in. knob spindle hole

⅜-in. top mounting screw hole

Bottom of mortise

1-in. holes to make key-hole mortise for thumb lever shaft (bottom hole drilled with ⅜-in. bit through to inside of door)

⅜-in. bottom mounting screw holes

⅜-in. bottom mounting screw holes (the distance of these holes from the top mounting screw hole varies depending on exterior trim style)

6 in.   6½ in.   7 in.

Template placed even with edge of door

A layout template made of ¼-in.-thick plywood quickly and accurately identifies the locations and the sizes of the holes that accommodate a mortise lockset's trim.

Baldwin template engage the latch plate mortise and automatically position the template perfectly. To avoid confusion and potentially expensive mistakes, mark the outside of the door first. Tap an awl firmly into the following five exterior layout holes:

1. Mark the top hole for the 1¼-in. key cylinder bore.

2. Skip one hole (that's for the interior dead-bolt knob) and mark the third hole from the top for the ⅝-in. knob spindle bore.

3. and 4. Mark each of the next two holes for 1-in. bores, but remember that they only penetrate half the door. These two holes are connected later with a chisel to form an elongated key hole for the thumb-lever shaft. (The lower hole is drilled through the remainder of the door, from the inside, with a ⅜-in. bit for the top mounting screw.)

5. Mark the location of the bottom mounting screw. That measurement was written down while checking the hardware. (More holes can be located in the plywood template for different exterior trim styles. The Baldwin template has seven holes, though I've rarely used more than the three shown in the drawing on the facing page.)

Replace the template ⅛ in. proud of the interior edge of the door, and mark the following four interior layout holes:

1. and 2. Skip the top key cylinder hole and mark the next two ¾-in. holes: the first hole for the dead-bolt thumb turn and the next hole for the knob spindle.

3. Skip the next hole (which is only drilled on the outside of the door for the top of the keyhole-shaped thumb-lever shaft), and mark the next hole (fifth hole from the top) for the top mounting screw, which from the inside of the door is bored ⅜ in. and completes the 1-in. bore started from the outside of the door.

4. Mark the appropriate ⅜-in. hole for the bottom mounting screw.

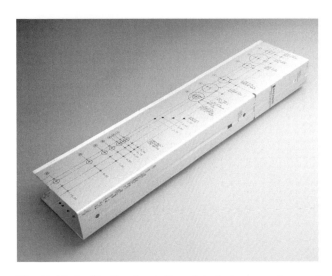

The Baldwin Mortise Lock Template is inexpensive and works perfectly. It has holes for every available trim design, and also includes hole locations for 2¾-in. backset locks.

Start all the through holes from one side and finish them from the other side to avoid tearout. Before slipping the lock case into the mortise, use a chisel to connect the two 1-in. holes for the thumb-lever shaft. The upper hole lies within the large pocket mortise, but the lower hole does not. Use a small sharp chisel to pare away the back of the lower hole until both key-hole bores are of equal depth. If the lower hole isn't deep enough, the thumb-lever shaft will rub and stick.

Assembling the Lock
Before inserting the lock case, the bevel must be adjusted to match the bevel of the door or the key cylinder will not lie flat against the exterior trim. There are three bevel adjustment screws, one at the top of the lock case, just behind the main latch plate, and two at the bottom of the lock case. Loosen all three screws just a little, hold the lock case flat against the face of the door and rotate the latch plate to match the bevel of the door, then tighten all three screws. This step is so important that Baldwin even suggests shimming the lock case as it's installed, otherwise as the cylinder is tightened into the case, the case may be pulled off center, and the cylinder won't sit flat on the exterior trim.

ADVANCED DOOR HANGING   **151**

Install the knob spindle before attaching the exterior trim plate. (Photo by Rich Ziegner.)

Thread in both handle mounting screws, but don't tighten them until the key cylinder is installed. Remember to slip the rosette base onto the lower mounting screw. (Photo by Rich Ziegner.)

Threading the key cylinder into the lock case is easier if the handle set can still be moved. Use a Schlage key to turn the cylinder because Baldwin keys are made of soft solid brass and sometimes they twist in the cylinder. (Photo by Rich Ziegner.)

Install the interior trim and knob. If the knob threads onto the spindle, tighten the knob until the latch hangs up in the case, then back off one turn and tighten both set screws. If the knob slides on to the spindle, push the knob hard against the trim while tightening the set screws. (Photo by Rich Ziegner.)

Once the lock case is placed in the pocket mortise, use the following steps to install the trim, referring to the photos on the facing page. First, insert the knob spindle into the lock case (see the top left photo). Most spindles now install from the inside of the door and are secured from the outside of the door by a small Phillips-head screw and lock washer. Before threading in the screw, paste it with glue or lock-tight, because it has a habit of backing out during use, and the screw can't be tightened later without removing the exterior trim.

Next, place the exterior handle set on the face of the door and thread in the top mounting screw just enough to secure the trim. Start threading the bottom mounting screw into the exterior trim as well, but be sure to slip the rosette base onto the bottom screw first (see the top right photo).

Now, insert a small shim through the key cylinder hole so that the lock case won't move as the cylinder is installed, then thread the key cylinder into the lock case until it's snug against the trim plate (see the bottom left photo). Use a Schlage key or small screwdriver to turn and tighten the cylinder. Baldwin keys are solid brass and twist off easily. Then drill pilot holes and install the two main latch plate screws, after which you can tighten both handle-mounting screws.

After tightening the handle screws, check that the key cylinder is snug against the exterior trim, then tighten the cylinder retaining screw (in the latch plate above the dead bolt) so that the key cylinder won't spin. Don't overtighten this screw or the lock won't operate smoothly. (Always check the operation of a Baldwin lock after tightening any screws. Turn the key in the cylinder and be sure the dead bolt is working smoothly. If it isn't, loosen one screw at a time, just slightly, until the culprit is discovered.)

Install the interior trim next. Insert the thumb turn shaft into the lock case with the thumb turn in a vertical position (if the dead bolt is retracted). Push or thread the interior knob onto the spindle, then install the interior trim screws (see the bottom right photo). Before tightening the two set screws in the base of the interior knob, be sure the knob isn't jiggling because it's too loose and isn't sticking because it's too tight.

Finally, install the finished armor front plate and inspect the operation of the lock completely, before installing the rosette over the bottom mounting screw. More adjustments may have to be made. Besides, it's not uncommon to forget something, and then the entire lock may have to be disassembled.

Installing the strike plate Because of the popularity of Q-lon type weatherstripping, it's best to locate the backset of the strike by measuring from the face of the jamb and not the door stop. Use the technique I described in Chapter 4 to locate the backset of the strike. The location of the strike can also be found by measuring from the interior face of the door to the face of the latch. Transfer that measurement onto the jamb, measuring from the face of the jamb; also transfer a mark onto the jamb $1\frac{1}{8}$ in. below the top of the latch on the door. To trace the outline of the strike, hold the strike on the jamb with the latch mark visible through the latch opening and the top of the strike $1\frac{1}{8}$ in. below the top of the latch.

I made a router template for ASA strikes—like the one supplied with a Baldwin mortise lock. I set the router's depth for the thickness of the strike plate, the dust bucket, and the security insert (a steel plate installed behind the strike plate that strengthens the strike plate). I also made a two-step template for ASA strikes and dust buckets for when I work in homes with numerous mortise locks. But it doesn't take that long to excavate the deep mortise for the dust bucket by drilling overlapping holes with a 1-in. spade bit.

After installing the strike, check the operation of the lock repeatedly. Throw and retract the dead bolt with the thumb turn; retract the latch and the dead bolt with the knob; and use the key, too, over and over, until there's no doubt that the lock is working properly.

Installing a Baldwin (ASA) Strike Plate

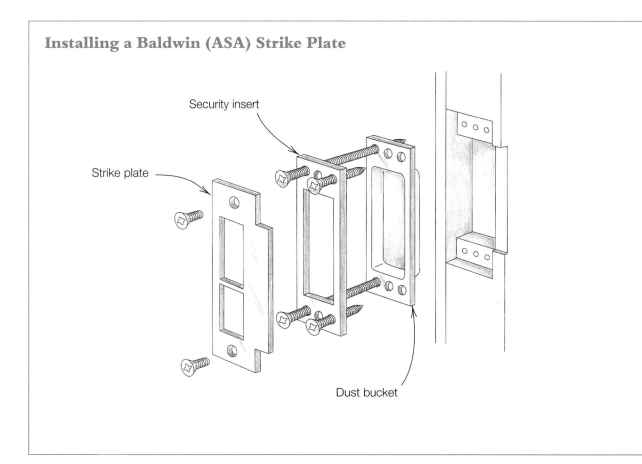

Installing Schlage Mortise Locks

In Chapter 2, I said that Baldwin mortise locks are like Cadillacs; Schlage mortise locks must be like Range Rovers. Schlage L locks aren't available in all the fancy trim designs, but with their utilitarian engineering, they can survive any terrain—hospitals, office buildings, and even schools.

As with any mortise lock, always check the handing first. The handing of a Schlage L lock can be changed in the field by pulling the latch tongues straight from the case and rotating them 180 degrees. There are no set screws securing the latch. The bevel on the top and bottom latch tongues must face toward the jamb, but the antifriction tongue in the center, which actuates the retraction of the latch, must be square to the jamb (see the drawing on the facing page).

The internal function of the lock has to be changed, too, so that when the door is locked the lever on the outside is inoperative and not the lever on the inside. On a Schlage lock changing this function is easy: The lock case doesn't even have to be opened, and only one screw has to be removed. Remove the catch screw (the only round-head screw in the lock case, located near the lower corner of the lock case, beneath the spindle holes) and replace the catch screw on the opposite side of the lock case. The catch screw should always be on the inside of the door for the lock to operate properly.

Bouvet Mortise Locks

Many companies have imitated Baldwin's lock case, and some have even added improvements. Identical to Baldwin's lock case in size and shape, Bouvet's case has the advantage of being reversible. Remove one screw at the rear of the latch tongue linkage (near the back of the lock case), then pull the latch tongues clear of the case, rotate them, and slip them back in. Replace the screw to secure the latch in the case.

A second set screw, located beneath the first, secures the adjustable function button on the front of the lock case. Remove this set screw, reverse the slotted function button, then replace the set screw. Bouvet supplies a chart to determine the position of the slotted function button, but the chart can be confusing. So before installing the lock case, check the functions to be sure the slotted button is in the proper position.

Mortise lock case

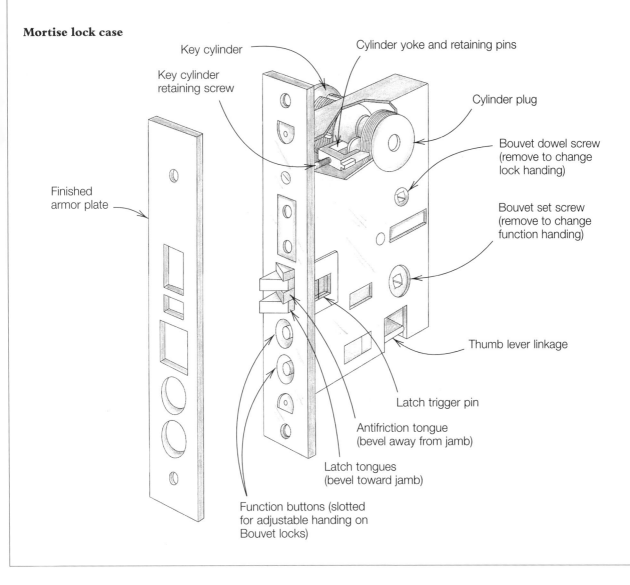

ADVANCED DOOR HANGING

Schlage mortise locks are unusually simple. I've stacked this one up in the order of installation. Each handle is mounted on a separate spindle with its own spring cage (foreground left side and inserted in bottom of lock case beneath cylinder). As with all mortise locks, always thread the cylinder into the lock case and insert the handle mounting posts before securing the lock case in the door.

Assembling a Schlage L lock Not only is the mortise for a Schlage L lock wider, taller, and deeper than typical residential lock cases (Schlage locks require a 1-in.-wide by 6¼-in.-tall mortise; other cases require a ¹⁵⁄₁₆-in.-wide by 6-in.-tall mortise), but the trim and parts for a Schlage L lock are different, too. Fortunately, the armor latch plate cover is the same size as most mortise locks, so the same router template can be used. Once the pocket and latch mortises are complete and all the cross bores have been drilled (see the sidebar on the facing page), assembling the lock is fast. Adjust the bevel, insert the lock case in the mortise, and use the following steps to finish the lock (see the drawing on p. 154).

First, thread the key cylinder into the case, then install the two latch screws and fasten the case to the door. Secure the cylinder by tightening the retaining screw closest to the outside of the door.

Next, insert a spindle through each side of the door into the hexagonal lever hub in the lock case, and slide a spring over each spindle (insert the short end of the spindle until the pin-stop rests against the hub).

Now, thread the two extension screw posts onto the exterior handle rosette, and slide one spring cage over the posts (with the arrows pointing in the direction the handle must rotate). Then install the exterior handle on the door slowly so that it engages the exterior spindle.

156 CHAPTER SIX

Making Perfect Lock Templates

For Schlage mortise locks and all other odd-layout locks, like Bouvet patio locks and Von Duprin panic hardware, I make special templates that wrap around the face of the door (similar to the Baldwin template shown in the photo on p. 151). The wooden back plate of the template is rabbeted to fit inside the latch-plate mortise, which ensures that the template will always be aligned precisely. I locate the holes in the clear plastic sides by testing the template against doors that are bored perfectly, so that no hole ever has to be enlarged.

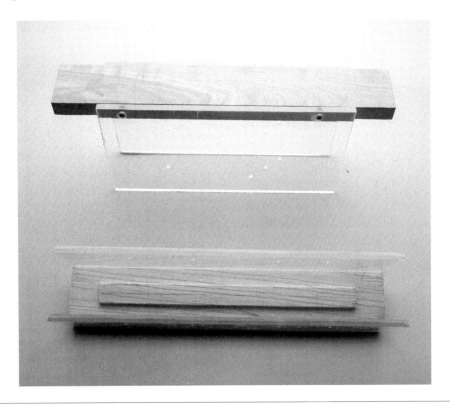

Slip the second spring cage over the extension screw posts that now penetrate the inside of the door. Then slip the circular threaded lever mount on top of the spring cage and tighten the two mounting screws.

Finally, place the interior rosette on top of the mounting plate and engage and then thread on the interior handle. Tighten the handle with the spanner wrench that's supplied with every lock, then install the armor latch plate.

Chapter 7
UNCOMMON DOORS & UNUSUAL HARDWARE

SIDELIGHTS AND LATCHES

ARCHED DOORS

DUTCH DOORS

DOUBLE-ACTING DOORS

INVISIBLE HINGES

SITE-BUILT BIFOLD DOORS

POCKET DOORS

The previous chapters of this book have concentrated on the fundamental skills that make the craft of hanging doors simple and enjoyable—even hanging a hollow-core interior door can be rewarding. But challenging projects are the pinnacle of the craft. Unfortunately, the exact techniques, the precise measurements, and the step-by-step setup we used two years ago to hang a double-acting door, an arched door, or a door on invisible hinges are difficult to recall. In this chapter I will review the techniques I use for uncommon doors so that even though a carpenter may only install one double-acting door every few years, every necessary detail can be recalled.

SIDELIGHTS AND LATCHES
Because they seem simple but mostly because they're rarely encountered by many carpenters, sidelights often present a challenge. Whether they're permanently attached to the jamb or installed on hinges like any swinging door, installing a narrow sidelight requires different scribing techniques and different plane setups than those used on a standard door.

Stationary Sidelights
Sidelights are most often installed in exterior openings, which usually include a sill or threshold (see Chapter 3). Stationary sidelights should be cut to fit tight to the sill, to prevent water infiltration, and they should be planed to allow a small gap between the jamb and the stiles and top rail, so that the fit matches that of adjacent doors. I

The new pair of doors required most of the opening's width after the original sliding patio door was removed from this exterior wall. Narrow-layout sidelights with 2½-in. stiles allowed more glass space in each sidelite.

always scribe and install sidelights in jambs with back-to-back mulls before scribing the doors. This is especially true for stationary sidelights: Because a stationary sidelight is nailed into the jamb, the fit of the active door is dependent upon the fit of the sidelight.

But before setting up to scribe the sidelights, be certain that all the top rails and muntin bars will be level and aligned. Occasionally, sidelights are manufactured differently from doors (the stiles of a narrow-layout sidelight can vary in width from 2 in. to 3 in.), and the top rails can be wider or even narrower than regular doors. Always check the sidelights against the doors. Measure from the top rail to various muntin bars and be sure the layouts are equal. If they aren't equal, correct the problem before scribing the doors by trimming the wider top rails.

Because the top rails must be scribed and cut perfectly, accurate shimming is crucial before scribing. Shimming stationary sidelights beside an oak sill or aluminum threshold can be difficult: The shims can't be driven perpendicular from the inside very far before they run into the sill. And stationary sidelights cannot be shimmed from the outside because the sidelight must fall below the sill or threshold in order to be scribed. It's easiest to shim doors against an oak sill with scrap pieces of plywood in varying thicknesses. Raise the doors enough so that final adjustments can be made with a small shim.

Scribe the stiles on a stationary sidelight the same as an active door, with scribes spread about 1/8 in. apart. Sidelights are easier to install if both sides are beveled.

Stationary sidelights don't require a bevel, but being accustomed to beveling doors, I bevel both stiles and scribe the stiles the same as a regular door, as shown in the top photo at left. Be sure to squeeze the scribes closed to trace the head of the jamb against the door. To scribe the sill, turn the scribes over so that only the pencil rides on the top of the sill—the bottom of the sidelight should be tight to the top of the sill.

Planing and cutting sidelights accurately can be a challenge, because even a slight inconsistency is apparent across a narrow head. Inexperienced door hangers should aim for a tighter fit at first, then carefully plane more material until an exact tolerance is achieved. Before installing a sidelight, seal all four edges with two coats of exterior sealer, especially the bottom, otherwise the sidelight will wick moisture from the sill and quickly begin to rot.

Unfortunately, a perfect fit can be ruined if precautions aren't taken while nailing the sidelight into the jamb. I use 8d finish nails to shim between the jamb and the stiles of a sidelight, before driving any nails. Once the sidelight is fastened, the nails are easy to pull and leave no marks—only a perfect margin between the jamb and the sidelight.

Active Sidelights

Sidelights beside an entry door are usually stationary, though sidelights beside French doors should be active because they can be screened, just like a window, which eliminates the need for ugly screen doors (see the bottom photo at left). Scribing an active sidelight is much different than scribing any other door. A sidelight is usually much narrower than a regular door, so the bevel must be steeper (see the drawing on the facing page), and because the bevel is steeper, the scribes must be spread farther apart to anticipate the extended width of the door caused by the longer bevel.

Normally I spread my scribes about 1/8 in. apart to accommodate the bevel and necessary gap on a door that measures wider than 2 ft. On 1-ft. 6-in. sidelights, I spread my scribes an additional 1/16 in. and increase the angle of the bevel; for narrow sidelights, I spread my

Sidelights are not limited to pairs of doors. These two standard-layout sidelights allow plenty of ventilation and eliminate the necessity of a screen on the center door.

The Bevel on an Active Sidelight

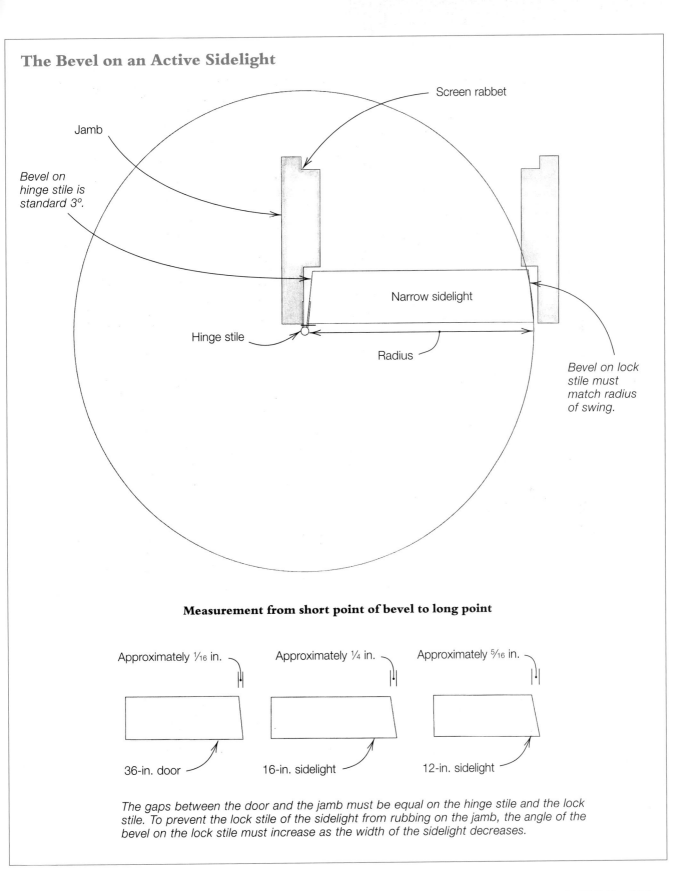

Once the casement latch is installed, close the door and mark the top and bottom of the latch plate. Most strikes are the same height as the latch and are meant to be installed level with the latch.

scribes a full ¼ in. apart. As on all doors, my aim with active sidelights is to achieve a perfect fit without having to take them off the hinges even once. But sidelights are easy to take down and put up, so plane a little at a time until the bevel clears the jamb and the gaps are consistent.

Latching Sidelights

Active sidelights require some type of latch or locking mechanism. If the sidelight has standard 4-in. stiles, then a regular 2⅜-in. backset dead bolt can be installed. But dead bolts aren't very attractive. Even simple thumb-turns that don't include an exterior key cylinder can be an eyesore; decorative surface bolts or casement latches are more pleasing.

To locate the exact backset of the strike, open the door and place the strike over the latch, with the tongue of the latch engaged snugly against the leading lip of the strike. Spread a pair of scribes from the face of the door to the back of the strike.

Use the scribes to transfer the backset of the strike onto the rabbet of the jamb, then align the strike with the two marks on the face of the jamb and trace the outline of the strike with a sharp pencil.

I use the same technique for all surface hardware, including casement latches and surface bolts. Begin by securing the latch to the lock stile of the sidelight. Mount the latch close to the edge of the door but not too close—don't allow the latch tongue to touch the trim on the back-to-back mullion (see Chapter 3) or the slide bolt to touch the casing. The tongue or bolt should extend as far as possible so that the latch securely engages the strike but doesn't damage the trim.

I use two cordless drills to install hardware. In one drill I chuck a Vix bit for drilling pilot holes (a Vix bit drills perfectly aligned pilot holes), and in the other drill I have a Phillips-head screwdriver tip. After positioning the latch on the door, drill one pilot hole and install one screw snugly, which temporarily secures the latch but still allows for minor adjustments. Then drill pilot holes for the remaining screws before installing them. Drilling the pilot hole with a Vix bit ensures that each screw will go where it's intended, without nudging the latch out of alignment, without stripping the head of the screw, without snapping off the screw, and without splitting the stile of the door.

Locating and installing the strike is often the most difficult part of installing any surface latch. The method I use, shown in the photos on these two pages, simplifies the process and ensures a precise fit every time. You can use the same approach for surface bolts, too. Attach the bolt first, then position the strike over the bolt to determine the backset of the strike. Then spread your scribes from the face of the door to the back of the strike, and use the scribes to transfer the backset of the strike onto the jamb.

Use a sharp chisel to impress the outline of the strike. On narrow sidelights where there isn't enough room to swing a hammer, drive the chisel with the side of the hammer.

Finish the mortise from the face of the jamb so that the strike sits in a clean mortise, then drill pilot holes and install the screws.

Installing a pair of arched doors isn't as difficult as it seems if the right steps are taken in proper order.

ARCHED DOORS

Installing arched doors is the crown of door hanging (see the photo at left). Fit properly, arched doors are a thing of beauty and a trick of the imagination—they seem so difficult to install but they're hung just like any other door. Using these techniques, even arched doors can be scribed and cut to fit perfectly the first time up on the hinges, and the only requirements are patience and precision.

The head of a flat jamb acts as a stop for a hinge template, and the top edge of a door represents the corresponding stop. But arched jambs are different: There's no flat jamb head to stop a template, and there's no corresponding point on the door to which a template can be registered. This is why I always cut hinge mortises in arched jambs first so that the hinge locations can be marked while the door is being scribed (see the bottom photo at left). I use an adjustable hinge template for arched jambs, and I remove the top hook from the template so that I can get the top hinge as close as possible to the spring line of the jamb. On large doors with very tall arches, I'll even mortise the top hinge by hand, so it's located right at the spring line. Whenever possible I would rather use a hinge template, because templates and routers are faster and more accurate.

Hinge the jamb first so that the hinge locations can be transferred to the door while you're scribing the doors.

Scribe the door. Squeeze the scribes closed for the head and spread the scribes a little more than ⅛ in. apart for the stiles.

164 CHAPTER SEVEN

Scribing an Arched Door

Once the jamb is mortised, set the doors in the opening and hook them to the jamb. Spend the necessary time shimming and preparing to scribe because there's no latitude on an arched door. Once an arched door is cut and mortised for hinges, the head gap should fit perfectly and never be too big. This is one instance where a pair of scribes is more useful than a pencil. A pair of scribes can be squeezed closed to equal the head gap of the door, as shown in the right photo on the facing page. Carpenters unfamiliar with arched doors should make this initial head-gap scribe tight; additional trimming and planing can be performed after the doors are swinging. When scribing the head of the jamb to the door, be careful to keep the scribes perpendicular to the jamb—don't allow the scribes to tip or the pencil line on the door won't be true to the jamb.

After scribing the head, spread the scribes to accommodate the bevel on the hinge and lock stiles (on this pair I'm only scribing hinge stiles). Don't worry about the odd intersection of the scribe lines where the hinge stile meets the head of the door. Though it appears the hinge-stile scribe will cut too far into the head-gap scribe, it all works out perfectly because of the bevel on the hinge stile. As I said earlier, while the doors are still hooked to the jamb, mark the locations of the hinges, as shown in the left photo below.

Cutting an Arched Door

As always, the first job with any door is to cut the bottom and the head. Surprisingly, a power door plane works extremely well on an arched head. The only difficult chore is keeping the sole of the plane plumb—if the plane tips even slightly, the cut will be ruined and the tops of the doors won't fit. Since it can be hard to see if the plane is cutting plumb, I wrap my left hand around the head of the plane until my fingers can feel the fence and the opposite face of the door, as shown in the center photo below.

Because most of the plane's sole never touches the door (only a few inches of the sole near the cutter are in contact with the door), the depth lever serves no purpose. Instead, the depth of cut is controlled by altering the amount of the cutter that's in contact with the door by carefully rocking the plane. For that reason, always begin with the plane inverted and work toward the opposite stile. Watch the cutter as it approaches the scribe line and take a small, even bite in as long a pass as possible. Make repeated passes with the plane, carefully

Mark the hinge locations carefully because they affect the head gap.

Plane the top of the door but leave the pencil line.

Sand the top of the door right to the line.

UNCOMMON DOORS & UNUSUAL HARDWARE

If the head gap is a little too tight in some spots, scribe a line on the face of the door then use a belt sander cautiously.

approaching the scribe line each time. Only within the final inches of the cut should the plane be held motor up, because the scribe line will be hidden by the plane's motor as the top of the far stile is trimmed. While planing the head of the door, be certain to leave the pencil line so that it can be used as a guide for final sanding, as shown in the right photo on p. 165.

The scribe line on the hinge stile intersects part of the arched head, which can cause some confusion. It may appear as though the hinge stile scribe is intersecting the head too far above the spring line. But the confusion is straightened out once the hinge stile is planed at a bevel. The short point of the bevel intersects the scribe line on the arch; but on the face of the door, the long point of the bevel is at the exact intersection of the spring line between the jamb leg and jamb head. Plane the hinge stile just like a rectangular door. Hold the plane steady and run it straight out of the door. Don't follow the arch as the plane enters or exists the door.

If the fit isn't perfect, use a pair of scribes to trace the line of the jamb onto the face of the door (see the photo at left), then sand or plane any areas that are too high. Minor trimming with the plane or a belt sander, to really perfect the fit, is sometimes possible without removing the door from the jamb. Easing the top edge of the door slightly will also cause the gap to appear more uniform.

DUTCH DOORS

Oddly enough, and unlike arched doors, installing a Dutch door successfully has more to do with the jamb than with the door. The hinge leg of a Dutch door jamb must be set perfectly straight, plumb, and rigid. The slightest deviation in the jamb will affect the fit of the doors, not only as they lie flat against the door stop but more importantly as they swing open. A concave bow near the center of a hinge jamb will cause a pair of Dutch doors to pinch and bind as they're swung open; conversely, a convex bow will cause the doors to spread as they're opened and can result in an unsightly large gap between the two open doors.

Setting a Dutch Door Jamb

Though it's preferable to plumb and straighten the trimmer on the hinge side of the door, remodels don't always allow that alternative, in which case the hinge jamb must be shimmed carefully and fastened thoroughly. For Dutch doors, I use long drywall screws rather than nails to secure the jamb. I use screws not only because the weight of a Dutch door will pull down on the top of a jamb (the upper door can cause the jamb to tweak at the scissor point between the two doors) but also because hinge screws can deflect a jamb. Sometimes the screws strike against the trimmer before they begin to thread; sometimes they don't want to thread into the trimmer at all but instead push the jamb away from the trimmer. If the hinge screws move the jamb even a little, fitting a pair of Dutch doors can be a difficult and even impossible task.

Because these jambs require more attention to detail, a logical order of installation can save a lot of wasted effort and time. Start by securing the jamb in the opening, but use only a few shims and nails. Then decide on the hinge locations and mortise for the hinges as soon as possible. The two top hinges and the bottom hinge are located the same as on a standard door, so a standard three-hinge template usually works fine (see the drawing on p. 168). I use a short single-hinge template to mortise for the top hinge on the lower door. Once the hinges are mortised, it's easy to be sure that there's solid backing and shims behind each hinge. But don't break off the shims yet. Use one hinge to trace the locations of the two inner hinge screws, which are closer to the center of the trimmer. Countersink and drill a hole between the two inner hinge screws, then drive in a long drywall screw. Check the jamb with a long straightedge to be certain the drywall screws aren't pulling the jamb too much. Because the shims haven't been broken off yet, they can still be adjusted. Try to correct even the slightest variation in the jamb so that hanging the doors will be easier.

Hanging Dutch Doors

If the jamb is set properly, hanging a pair of Dutch doors is easy. If I'm alone on a job, I often join the two doors temporarily, so that it's easier to scribe them. But I prefer to handle one door at a time and have a helper hold the upper door against the opening until I've scribed it. As I said earlier, I use a standard 6-ft. 8-in. three-hinge template for most Dutch doors, so hinging the top door is simple. Just let the long template hang way over the bottom of the door.

Swing and close the top door before scribing the bottom door, because the upper door is like the head jamb for the lower door. If a rabbeted shelf is being added to the top of the lower door, the scribes must be spread the thickness of the shelf plus ⅛ in. for the doors to clear each other. The door opening in the photo at right was in an exterior wall, so I installed a shelf below the rabbeted joint between the two doors (see the detail in the drawing on p. 168). While scribing the lower door, locate the position of both hinges on the door. Both lower hinges should be mortised with a single-hinge template because a three-hinge template won't work on either one—the hinge layout is completely different.

Dutch doors present a special challenge. The hinge leg of the jamb must be strong and straight. The back-to-back mull for this Dutch door required special reinforcement.

Dutch Door Dimensions

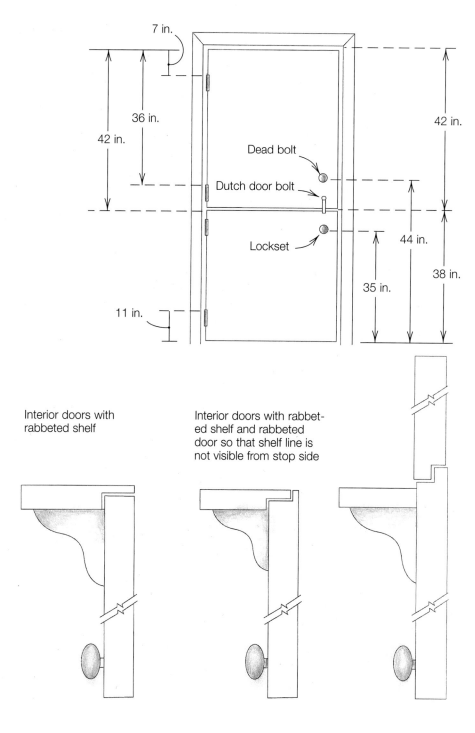

Jamb Reinforcement

Not long ago, my brother Larry and I were asked to install a Dutch door in a client's newly remodeled entryway. Aware that Dutch door jambs were problematic, we were doubly concerned about the client's flanking-sidelight design. The homeowner wanted both sidelights to be active, though we tried to talk him into accepting a stationary sidelight behind the Dutch doors. But design took priority over pragmatism, and we worried about that narrow back-to-back jamb supporting a Dutch door. Whenever in doubt, Larry's answer has always been steel, and it was in this case, too. We reinforced the back-to-back hinge jamb for the pair of Dutch doors so that it couldn't even move a little bit, and we carefully kept the steel away from the hinge-screw locations, too.

Jamb reinforcement for a Dutch door

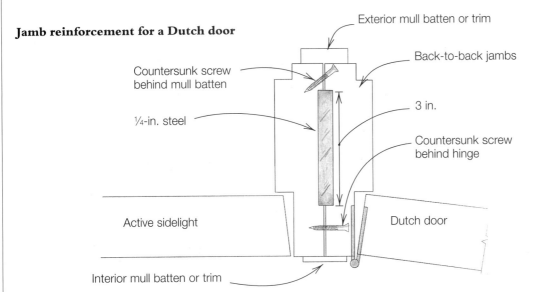

Use a tablesaw and dado blade or a circular saw to cut the shallow 1/4-in.-deep by 3-in.-wide mortise. Screw and glue the two jamb legs together, like any back-to-back mullion, before assembling the frame.

If the hinge jamb is perfectly straight, both doors should fit uniformly in all positions. When boring for the lockset, locate the handle far enough below the shelf to allow room for a hand! Then install *and retract* the Dutch-door slide bolt (to be sure there's enough room to retract the bolt) before locating the dead bolt position.

DOUBLE-ACTING DOORS

For most carpenters, the installation of a double-acting door is like an encounter of the third kind: Maybe once every few years or more infrequently does a double-acting door materialize, and then it's always a struggle to remember the steps and measurements that worked the last time. For freshman or even sophomore door hangers, double-acting doors can be a daunting challenge, not the least because the instructions are always

UNCOMMON DOORS & UNUSUAL HARDWARE **169**

Double-acting doors are the perfect answer for an opening between a dinning room and kitchen, but installing them requires patience.

minimal and often confusing. The information in this section is meant as a guide for all types of double-acting hardware, because they all share common requirements.

Installing the Jamb Hardware

Always install the jamb bracket for a double-acting door before scribing the door, so that the distance between the bottom of the door (which is also the bottom of the double-acting hinge, though not the hinge base plate) and the floor can be scribed from a fixed point, rather than measured. Rely more on a pair of scribes to minimize errors and maximize the certainty of the fit.

The jamb bracket is normally installed in the center of the jamb, though I have set double-acting doors flush with one side of the jamb. In some kitchens, cabinets and double-acting doors conflict, and the less the door projects into the kitchen the better. The situation can also be reversed for a small dinning room, where the door might interfere with furniture layout.

The height of the jamb bracket is critical. On hard-surface floors, the top of the bracket must be no lower than flush with the finished floor in order to install and remove the hinge. On carpeted floors, the bracket can be installed a little lower than the projected top of the carpet because the bottom of the hinge (which will be the bottom of the door) clears the top of the bracket by ½ in. If a hard-surface floor is already installed, then set the bracket right on the floor, but if the floor isn't installed, a spacer must be cut to support the bottom of the bracket (see the top left photo on p. 172).

Installation instructions that accompany these hinges vary. Some instructions recommend installing the jamb bracket on the surface of the jamb, while some recommend mortising the bracket flush with the jamb. Because of the variance in installation instructions, and because I prefer not reading the instructions every time I install a double-acting door, I've adopted a set of layout measurements that work every time and result in a tighter, cleaner fit (see the drawing on the facing page). And I always mortise the bracket flush with the jamb.

After tracing the outline of the bracket on the jamb, I use a small cordless saw to cut the top of the mortise. I dislike chiseling against the grain, especially when there's so little access for a chisel from below the mortise. I set the depth of my saw blade to the thickness of the bracket. Using the saw speeds up the mortising process significantly, because chisel work can be accomplished laterally and down. Once the bottom bracket is finished, I move to the top pivot. I cut the top pivot mortise completely with a spade bit, which is easy, but the layout is critical (see the bottom photos on p. 172).

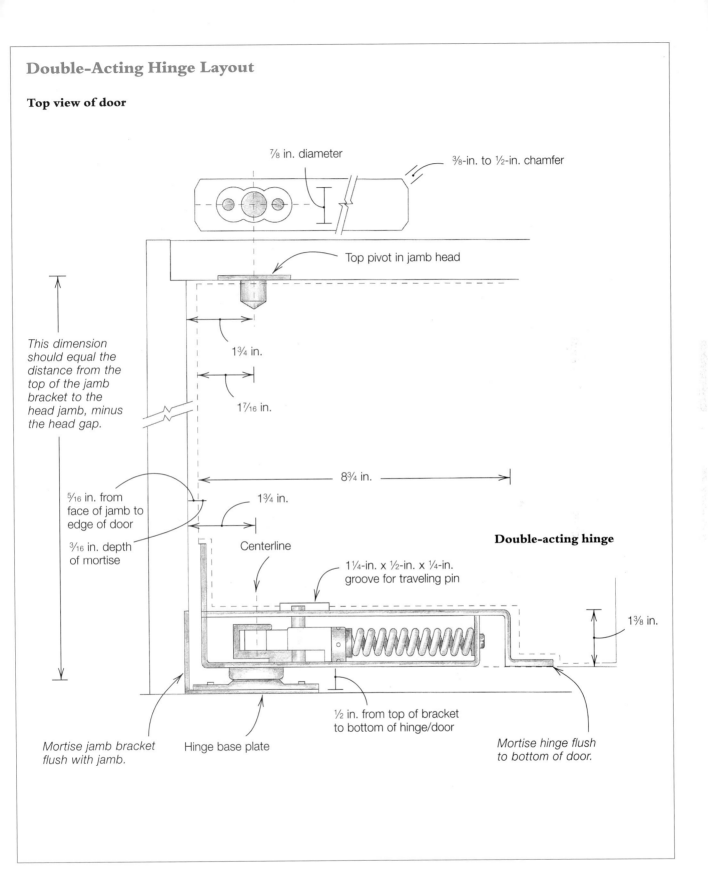

If a hard-surface floor hasn't been installed or if the floor will be carpeted, use a spacer to support the bottom of the bracket. This spacer is high enough to clear 1½ in. of mud bed and terra cotta pavers. Glue the spacer to the subfloor so that it won't ever move.

Scribing the Door

Rather than scribing the door for a large ⅜-in. gap, as some installation instructions suggest, I try to make the size of the hinge and strike gap a little smaller. I've hung double-acting doors as tight as ¼ in., but 5/16 in. is a safe compromise between the instruction sheet and aesthetics. There's no bevel on either stile because the door swings in both directions, so the gap will equal the spread between the scribes. The head of the door is always scribed ⅛ in. from the jamb, which allows enough clearance, because the top pivots are mortised flush.

The size of the scribe on the bottom of the door is extremely important because the bottom of the door represents the bottom of the hinge, which in turn determines the fit between the top of the jamb bracket and the head of the jamb (see the left photo on the facing page). The cut must be exact: When it's time to install

Before scribing the door, install the top pivot. Hardware varies, so follow the installation instructions to find the distance from the jamb to the center of the pivot. Draw lines through that center mark, parallel and perpendicular to the hinge jamb.

Center the top pivot over the lines. The screw holes help to center the pivot. Carefully trace the location of each screw hole. The right size spade bit will cut the mortise and the pilot holes at once. For this pivot I use a ⅞-in. bit.

the door, the hinge must slip easily over the jamb bracket, without being too loose or too tight.

Planing the Door

Rather than cutting the top and bottom of the door first, like a standard door, plane the stiles of a double-acting door first (see the right photo below). After planing the stile square to the scribe line, tip the plane at a 45 degree angle and cut a ⅜-in. to ½-in. chamfer on both edges of each stile. A belt sander can be used to ease the edges further and even to round the door to a smooth radius (most instructions recommend a 1½-in. radius on the hinge stile). But a radius isn't always the best look for every house. Besides, the appearance of the door is improved if the hinge stile and the lock stile match.

Another alternative to planing a chamfer is routing a ½-in. or ¾-in. bullnose on each edge of both stiles. Not only is using a router faster than planing and sanding, but the cut is always consistent. The point here is that a double-acting door can be planed many different ways and doesn't have to have a 1½-in.-radius hinge stile and a square-cut lock stile. A double-acting door can be cut to resemble a standard door without compromising the function of the hardware.

Installing the Bottom Hinge

Even though it may look difficult, installing the bottom hinge is easily accomplished as long as the right steps are made at the proper time. Always begin by carefully laying out the saw cut in the bottom corner of the door (see the left photo on p. 174). In order to bring the door closer to the jamb, the back of the hinge must be mortised deeper than flush in the stile of the door, but the bottom of the hinge should be flush with the bottom of the door.

Mark the door ½ in. above the top of the jamb bracket. Spread the scribes from the floor to that mark, then drag the scribes across the bottom of the door.

Unlike a standard door where the top and bottom are cut immediately, plane the stiles of a double-acting door first. Use the plane to cut a ⅜-in. to ½-in. chamfer on both edges of each stile, or use a router and ⅜-in. bullnose bit to cut a rounded edge.

It's easier to finish the hinge preparation and mortise for the hinge supports with the door standing on edge (see the right photo below). A sharp chisel will cut the mortises, but be careful chiseling the stile near the edge of the door. A router is much safer and cuts a cleaner, exactly proportioned mortise.

Before mounting the hardware, be sure to mortise for the traveling pin in the bottom of the hinge (see the left photo on the facing page). The movement of the pin can't be inhibited or the hinge won't operate properly. After mortising for the traveling pin, be sure to drill pilot holes for the mounting screws (see the right photo on the facing page).

Installing the Top Pivot

Because the pivot in the jamb head is centered 1¾ in. from the jamb leg, the pivot in the top of the door must be centered 1⁷⁄₁₆ in. from the back edge of the door to allow for a ⁵⁄₁₆-in. gap between the hinge stile and the jamb. But before mortising the pivot into the door, check that the pivot in the jamb is centered at exactly 1¾ in. If it's not, then adjust the position of the pivot in the door accordingly. Remember, the location of the pivot in the door determines the gap, and because it's difficult to make minor adjustments in the position of the top pivot, mortise for the pivot very carefully (once the mortise is cut and the screws are driven into the door, moving the pivot incrementally is frustrating). Use

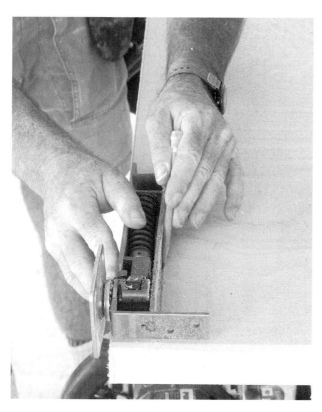

Hold the hinge ⅛ in. in from the stile, but keep the bottom of the hinge flush with the bottom of the door, then trace the outline of the hinge. To prevent tearout, score the horizontal line before cutting out the notch with a circular saw. Cut or plane the top of the door at the same time.

Stand the door on edge and trace the outline of both hinge supports. The mortises don't have to fit tight, but the depth is important. If the back of the hinge is let too far into the door, the door will be too close to the hinge jamb; in reverse the door will be too close to the strike jamb. I use a standard latch template with my router and set the depth carefully.

a ⅞-in. spade bit to cut the mortise, just like the jamb pivot, but before mounting the pivot in the door, drill a ½-in. hole in the center of the mortise for the jamb-pivot stud.

Hanging and Adjusting the Door

Stand the door near the opening, perpendicular to the jamb, and lift the door so that the top pivot stud engages the mating plate in the door, then slide the hinge over the bottom bracket (see the photo on p. 170). Slowly swing the door closed. The hinge will slide on the bottom bracket, but don't let it slide off. Align the holes in the hinge plate with the threaded holes in the bracket, then install the four machine screws to secure the hinge. If the door is unusually heavy, the tension on the hinge spring can be increased by rotating the collar at the back of the spring, though I rarely have to do this. After adjusting the fit of the door and tightening the retaining screws (see the top photo on p. 176), install the finished trim plates. Attach the face plates first with the finish wood screws, then attach the rear cover plate with the machine screw (see the bottom photo on p. 176). Three of the screws can be driven into pilot holes drilled in the door, but there's only air behind the lower rear hole. Often I flatten the rear plate slightly before installing it so that it follows the contour of the door.

A small groove must be mortised for the traveling pin that protrudes from the bottom of the hinge (see pencil), or the hinge won't swing smoothly and the door may not close completely. Mark the location of the pin on the door and use a ½-in. spade bit to drill a series of overlapping holes about 1¼ in. long and ¼ in. deep.

Be sure to drill pilot holes for all three mounting screws (the third screw is slightly hidden in the bend of the hinge). This isn't the moment to split the bottom lock rail inside the door.

UNCOMMON DOORS & UNUSUAL HARDWARE

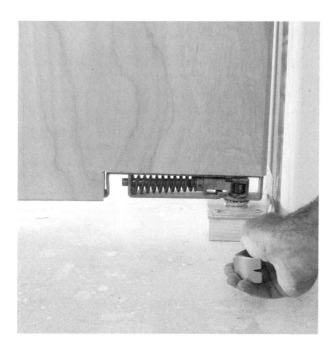

Make fine adjustments with a small prybar so that the screws won't cross thread. The position of the hinge on the plate is adjustable, so don't tighten the screws down until after making final adjustments. The hinge gap should be large, but the head gap and strike gap should be almost the same size as a standard door.

Attach the face plates first, before the rear cover plate. Three of the screws can be driven into pilot holes drilled in the door, but there's only air behind the lower rear hole. Cut off the last screw and set it in the lower hole with a small dab of silicone or glue.

INVISIBLE HINGES

The only way I know of to really make a door invisible, apart from waving a wand, is by using Soss hinges (see the top photos on the facing page). At first glance, installing Soss hinges might seem difficult, but by using the techniques I've already discussed, installing these tricky hinges can be almost as easy as hanging a standard door on butt hinges if each step is taken in the proper order. The first step is ordering the right door. Soss hinges are too thick to install in most doors because after mortising for the hinge there isn't enough wood left in the stile to support the hinge screws, especially for a heavy solid-core door. For flush doors, I've found the best alternative is to order a special door with 2½-in.-wide stiles (see Chapter 1).

Setting the Jamb and Backing Up the Hinges

Not only do Soss hinges require a special-layout door, but they also pose a problem for standard jambs. Because the shoulder of most Soss hinges is ⅜ in. thick, there's less than ⅜ in. of jamb material remaining after mortising for the shallow portion of a Soss hinge, and that's not enough wood to secure the hinge screws. For that reason, it's best not to shim a jamb that's getting Soss hinges, and instead to pick a clean and straight trimmer, set it perfectly plumb, then nail the jamb tightly against the trimmer. What's best isn't always possible, and if shims have to be used, try to use solid material rather than shim shingles. Slip different thicknesses of plywood between the trimmer and the back of the jamb until the space is filled. If the framing hasn't been finished, use 1½-in.-thick jambs. The thicker jamb material offers more purchase for both the deeply mortised hinges and the drywall taping where the jamb must be flush to the finished wall.

Soss hinges are also installed as little as ¼ in. from the edge of the jamb, which often leaves only drywall behind the hinge screws. Provide adequate backing for the hinge screws by installing a solid piece of wood behind each hinge, and secure the backing tightly to the jamb and the trimmer. Use a 1½-in. piece of clear wood for backing, preferably hardwood, and cut the backing long enough to span each hinge and yet still leave room enough for screws (see the bottom photo on the facing page).

Soss hinges can be installed in any direction, parallel or perpendicular to the door. The photo above left shows the door closed, with a reveal from the ceiling that matches the molding in the adjacent paneled wall. The photo above right shows the door open and the concealed pantry beyond the book-matched panelling.

Because most invisible doors are hung on jambs that are flush with, and also hidden in, the finished wall, the backing and the hinges must be installed before the drywall corner beading or L-metal is attached. After the hinges are installed and the jamb secured so that it won't move, the backing can be floated with drywall tape and mud and the wall finished.

Laying Out the Hinge Locations

Soss hinges are laid out differently than standard hinges. The manufacturer recommends using three hinges on solid-core doors and spacing the center hinge halfway between the top hinge and the center of the door, rather than centered between the top and bottom hinges, like a standard door. When I hang more than one door with Soss hinges, I use a story pole so that all the hinges are

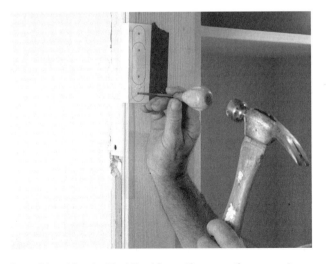

Install backing behind the hinge, then use the paper template from the manufacturer to lay out the hinge location. Tape the template to the jamb and use an awl to strike the center of each hole, so the spade bit won't wander.

UNCOMMON DOORS & UNUSUAL HARDWARE

A piece of masking tape works well as a stop on the spade bit. The depth of the top and bottom holes must be exactly the thickness of the hinge, and both holes must be perfectly square to the jamb, with flat bottoms.

Afer all the holes are drilled, pare the sides until each hinge fits easily in and out of the mortise. Chiseling is made easier because the jamb is backed up by a rigid piece of hardwood.

in the same location on every door and jamb. Any 1x2 works fine for a story pole, as long as it's almost the height of the door. Lay out the hinge locations on the story pole first, then use the pole to transfer those locations to the jamb. To allow proper head gap, mark each hinge location $\frac{1}{16}$ in. down from the top of the stick. When transferring the hinge locations to the door, keep the stick flush with the top of the door.

Mortising for the Hinges

A paper template is supplied by the manufacturer, and it works well for laying out the hinge mortises. First, fold the paper template along the line that represents the jamb, then tape the template at each hinge location to mark the center of each hole. For drilling flat-bottomed holes, a Forstner bit works better than a spade bit, but it's easier to stop a spade bit from wandering. In either case, use a piece of tape on the bit as a guide to determine when to stop drilling the top and bottom holes (see the photos above). The center holes can be drilled without as much care, but the shoulders of all four holes must be straight because they form the sides of the hinge mortise.

I prefer using a router and template for mortising Soss hinges because a router and template are easier, faster,

and more accurate. Even a simple one-step template works for routing the initial outline and depth of the hinge, but a two-step template (see Chapter 6) simplifies lay out, speeds up mortising, and ensures accuracy. Templates for Soss hinges, meant for use with $\frac{5}{8}$-in.-diameter template guides, are available from the hinge manufacturer. I made a special template for the $\frac{3}{4}$-in. template guide on my Bosch plunge router (see the left photo on the facing page). I also attached a stop to the back of the template to simplify and reduce layout time.

Swinging the Door

There isn't very much adjustment in a Soss hinge, which is why the hinge screws shouldn't be tightened too much. A small amount of adjustment can be gained by moving the hinges slightly: Tightening the screws will draw the hinges into the door or jamb; loosening the screws will let the hinges out of the jamb or door, so the door can be coaxed a little toward the hinge jamb or the strike jamb. Though it's not uncommon to have a door on Soss hinges fit the first time up, it's more the exception than the rule. These doors should fit tight all the way around—the $\frac{1}{16}$-in. gap at the head should be matched at the hinge and strike jamb, which is why I

A two-step template is the best way to mortise for Soss hinges. A plunge router with a long bit will reach to the bottom of the deep pocket and can then be dialed in perfectly for the crucial depth of the shallower steps.

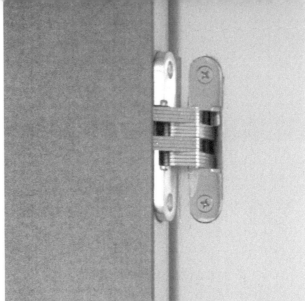

Be sure that the hinges slide easily in and out of the mortises, then secure the hinges in the jamb first. Unlike a standard door with butt hinges, with Soss hinges, it is easier to insert the bottom hinge partially into the door first, then insert the top hinge.

initially scribe these doors tight. I'd rather take the door down for one more pass with a plane than have an invisible door that's not invisible.

SITE-BUILT BIFOLD DOORS

Bifold doors don't always come in a cardboard box, and they aren't always just for closets (see Chapter 4). There are other applications for these space-saving doors. In a wide opening that separates a dining room from a living room or den, between a bedroom and a sitting room, or between a master bedroom and bathroom, bifold doors can make a compelling aesthetic statement. I work for several architects and designers who appreciate the benefits of bifold doors, and they never cease to come up with new applications that are practical, attractive, and challenging. Because the hardware is unusual and the layouts are uncommon, custom bifold openings are a trick of the eye—they look more complex than they really are. Being familiar with the hardware is the only prerequisite to a successful job.

Custom bifold doors look a lot more complicated than they really are. These tall doors, mounted with a combination of pivots and butt hinges, are no more difficult to install than standard doors, if you figure each folding door as a single door.

Bifold and Pocket Door Hardware Dimensions

Type 1

- Bifold door each weighing up to 50 lbs.
- Multi-fold doors each weighing up to 30 lbs.
- Pocket doors weighing up to 125 lbs. (up to 150 lbs. with ball-bearing carrier)

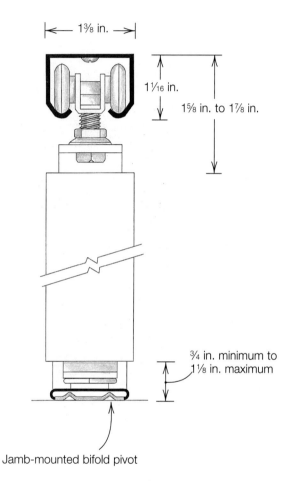

Jamb-mounted bifold pivot

Track interchanges with Hagar 9600, Stanley 1750, Johnson 111, Grant 7000, Lawrence HD 620/640.

Type 2

- Bifold doors each weighing up to 75 lbs.
- Multi-fold doors each weighing up to 125 lbs.
- Pocket doors weighing up to 250 lbs.

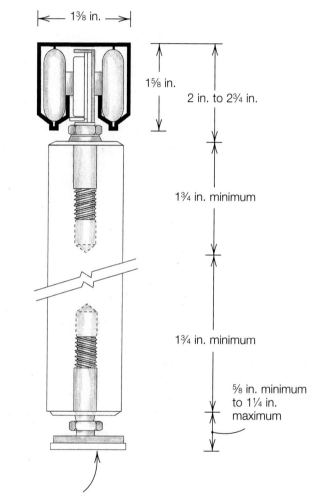

Heavy-duty floor-mounted bifold pivot shown mounted on bottom surface of door

Track interchanges with Hagar 9800; Stanley 2916-R, 2711, 2700; Grant 1260; Lawrence 570.

Type 3

- Bifold or multi-fold doors each weighing up to 275 lbs.
- Bifold or multi-fold doors each weighing up to 275 lbs.
- Multi-door pocket openings with doors weighing up to 275 lbs. each

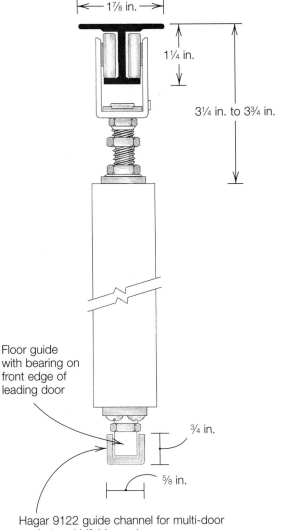

Floor guide with bearing on front edge of leading door

Hagar 9122 guide channel for multi-door pocket and bifold openings

Track interchanges with Grant 1210/1230, Hagar 9100.

Choosing the Track

Though the term bifold is commonly applied to all folding doors, the label is only applicable to openings with two doors; when more than two doors are used, the opening is called a multiple fold (multi-fold). One factor that partially determines the size of the track (and the type of track governs the height of the header) is the number of doors installed on the track—whether the doorway is a bifold or multi-fold. The other important factor is the weight of each door, which can vary according to the door size, construction, and composition. Solid-core doors are considered heavy (about 5 lbs. per square foot), but stile-and-rail doors with MDF panels weigh about the same. French doors are lighter, but ¼-in. glass weighs about 4 lbs. per square foot. I guess at the weight of most doors by using the 5 lb. per square foot rule. If the anticipated weight of each door even approaches the maximum rating of the track, I use the next size track. If there's any doubt, I weigh the door. I once placed a 6-ft. by 9-ft. French, dual-glazed pocket door on a scale and came up with 282 lbs. I used four wheel carriers on the Hagar 9110 track.

As shown in the bottom of the drawing on p. 182, folding doors can be designed in different multiples. When an odd number of doors is hinged together, the center doors don't pivot but instead swing off the previous door, much like an ordinary door. This design is useful for closing off a wide opening, yet still allows an easy-access doorway without the need to open the entire set of doors, which is helpful for separating a dining area.

Installing the Hardware

The track is always the first thing to install, and it can't be simply screwed to the header. The height of the header depends on the type of track (see the drawings at left). Often the header must be furred down to accommodate the track dimensions, and at the very least, the bottom of the header must be shimmed so that it's perfectly level and absolutely straight before the track is attached. I prefer to treat bifold openings like pocket doors, so I install the track before the drywall to eliminate any possibility of future drywall patching.

Before installing the track, the pivot location must be chosen. Bifold doors can pivot in the center of the jamb or hang flush with either face of the wall. Occasionally, I'm instructed to hang the doors flush with one side of

Bifold Door Designs

Bifold (two door)

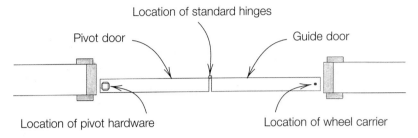

Four door (multi-fold)

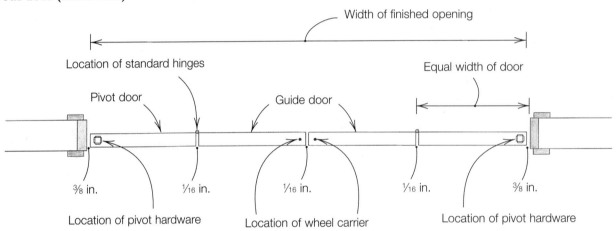

Six door (multi-fold)

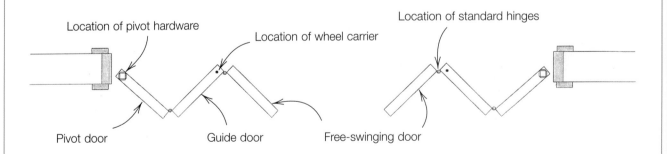

the wall so that the doors will consume as little space in the room as possible—installed flush with the outside face of the wall, much of the width of the doors remains in the opening, especially in a wide jamb. I've even installed bifolds in jambs that were exactly the width of the opened doors, so that the jamb and the opened doors were flush.

Once the track is installed, the bottom pivot can be seated, as shown in the photo at right. Unlike the jamb bracket on a double-acting door, the bottom pivot plate for a bifold door can be dropped beneath the finished surface of the floor. Doing so eliminates an unsightly large gap between the bottom of the doors and the floor. Determine the proper height of the bottom pivot plate by mocking up the plate and the pivot stud that will be mortised into the bottom of the door (see the drawing below). Bury the plate in the floor enough to absorb the thickness of the mounting plate and the lock nut, plus a little extra. Cut a plywood or hardwood block to support the bottom pivot, then glue and screw the block to the subfloor before installing the pivot.

The bottom pivot stud seats into a brass retainer in the bottom mounting plate. Mocking up these two pieces will determine how much the mounting plate should be recessed in the floor so that the gap between the bottom of the doors and the finished floor isn't too big.

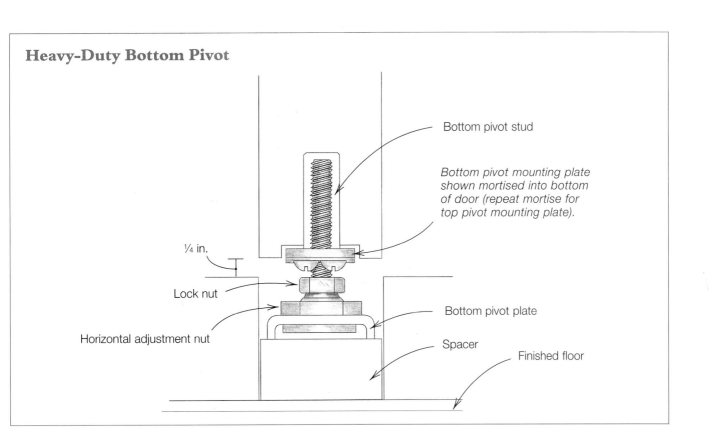

Heavy-Duty Bottom Pivot

UNCOMMON DOORS & UNUSUAL HARDWARE

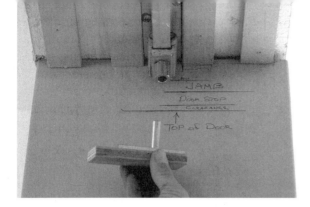

The top pivot attaches directly to the track. Notice the layout for the jamb and the door stop (which covers the wheel carriers) and the projected clearance between the door stop and the top of the doors.

Thankfully, the top pivot is a lot easier to install because it mounts directly to the track (see the photo above). The top pivot stud mates with the top pivot sleeve. The sleeve threads through the mounting plate and tightens against the track. Loosening the sleeve allows the pivot to move so that the doors can be adjusted. The large nut on the bottom pivot also provides for horizontal adjustment.

Hanging the Doors

Once the top pivot is installed, the line at the top of the doors can be located on the jamb. The working height at which the doors are initially cut should be slightly taller than the desired finished size to allow room for final scribing and trimming. To be safe, cut each door to the proper working height individually, rather than cutting all the doors at one time. Measure the first door at the jamb leg and measure each succeeding door from the preceding door.

Planing the pivot stiles isn't necessary if the jamb legs have been set perfectly plumb and straight. However, planing is necessary to make all the doors the same width. To determine the average door width, measure the opening, subtract for all the gaps (pivot gaps, hinge gaps, and strike gap), then divide that sum by the amount of the doors (see the drawing on p. 182). Usually the gaps end up slightly larger than estimated, with the result that the two center doors have to be planed just a little.

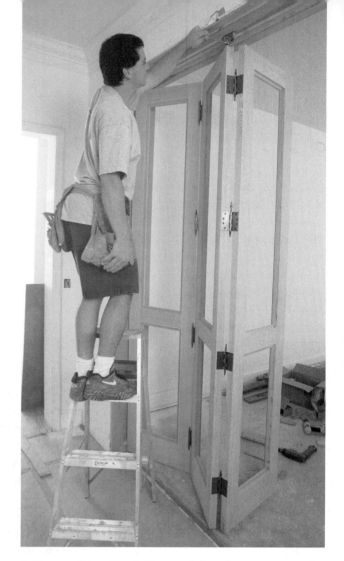

Before scribing the bottoms of the doors, engage the center doors in the wheel carriers and adjust the head gap by raising or lowering the carriers. The doors will certainly sag before engaging the carriers, which ensures that some of the weight will be on the wheels and the doors will slide smoothly.

After planing the stiles on the pivot door, mortise for the top and bottom pivot plates so that the top and bottom gaps can be reduced as much as possible. Otherwise the door must be an additional ¼ in. beneath the door stop and an additional ¼ in. above the floor to allow for the thickness of the mounting plates. Once again, a router and template will ensure that each plate is set at a uniform depth, so making a quick template for these mortises is the best thing to do. Even more important, the same template with a stop guide will help locate the wheel-carrier mounting plates on all the inner doors at exactly the same distance from the edge of each door, thus preventing the doors from binding as they move along the track.

Once the doors are swinging, adjust the top gap and the hinge gaps (see the right photo on the facing page). Because all the doors are joined to each other with standard hinges, the hinges can be adjusted as described previously. After making final adjustments, scribe all the doors to the floor at one time. Before removing any of the doors, scribe the meeting stiles of the center pair, too, so that all planing and cutting can be accomplished at one time.

Hanging Exterior Bifold Doors

A small number of manufacturers now offer large sliding and bifold door units that open an entire wall of a house to the outdoors. These doors are growing in popularity, particularly in warmer climates, and though some of the premanufactured units are expensive, it isn't that difficult to build a multi-folding exterior door unit with standard doors and hardware—but the planning stages are critical, because the doors must operate smoothly and be weatherstripped, too.

Exterior bifold doors are a challenge to design and build, but the reward is opening an entire wall of a home to the outdoors.

Standard pivoting bifold doors are nearly impossible to seal against the weather. The pivot points at the top and bottom of a pivot door, where the door moves in both directions, can't be sealed securely without interfering with the function of the hardware. But hinged bifolds can be weatherstripped, much like a standard door and frame. The drawing on p. 186 is representative of an opening I worked on that had 3-ft. by 9-ft. doors. To be certain the track was strong enough for the doors, I used the 9100 series Hagar track, but because of the long length of the wheel-carrier bolt, a two-step, 1-in.-thick door stop was necessary. For 6-ft. 8-in. doors, a lighter-gauge track can be used, which only requires a ½-in. to ¾-in. stop.

For this opening, I built up the exterior head jamb from two pieces, with the second piece kerfed for foam weatherstripping. The doors slide effortlessly against the slick, almost lubricated vinyl skin of the weatherstripping. Because the exterior jamb hangs lower than the tops of the doors, so that the weatherstripping will con-

tact the doors, I cut a small notch in the leading edge of the guide door, which allows that door to clear the head jamb as it travels along the track. The no-molding design of this plastered home created additional challenges, but in another house, this type of job can be accomplished with less effort by using exterior and interior trim.

Though the architect assured me this design would work, I've learned to trust doors and not drawings, so I mocked up a model of the opening as shown in the photo on p. 187. I also wanted to be certain that the calculated door widths would be correct before ordering the custom doors. The doors were expensive and I couldn't afford to risk an error. As shown in the drawing on p. 186, the guide door must be wider than the pivot door because it receives a wheel carrier that engages the track. The additional width is defined by the sum of the door thickness (1¾ in.) and the throw of the hinge (½ in.). I used a typical carpenter's approach to calculate the width of the doors in this opening. I di-

UNCOMMON DOORS & UNUSUAL HARDWARE

Exterior Bifolds in Bullnose Walls

Side view

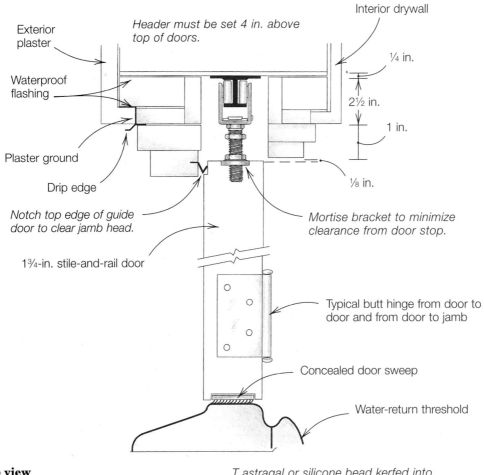

Top view

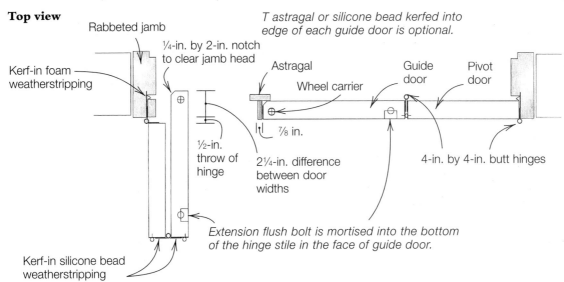

vided the opening by 4, and came up with 35½ in. for each door. The difference in width between the guide door and pivot door equals 2¼ in., which, divided in half equals 1⅛ in. I added that amount to the guide door and subtracted the same amount from the pivot door.

To seal the gaps between the pivot-door and guide-door hinges, I kerfed the edge of each door for silicone bead weatherstripping (see Chapter 8) before installing the hinges. I also kerfed the leading edges of the guide doors, though at the last moment I installed an astragal for better security when the doors are closed and locked by extension flush bolts mortised into the bottom face of both guide doors.

Mocking up an unusual exterior bifold design allowed me to calculate the size of the doors, test the weatherstripping, and ensure that the doors would operate smoothly.

POCKET DOORS

Like bifold doors, pocket doors fit into two distinct categories: manufactured frames and frames constructed entirely on the job site for custom openings. First I'll walk through manufactured frames, which are probably used 99% of the time. I'll explain the basics and the add-ons that make these frames easy to tailor for almost every need. I'll also describe a good step-by-step process for installing pocket-door frames. Then I'll step into the more complicated subject of custom pocket-door openings.

Manufactured Frames

Manufactured pocket-door frames are available in two common styles: "universal" frames, also known as pocket-door "kits," and ready-to-install built-up frames. I've used both types of frames extensively and have seen advantages and disadvantages to each, but one aspect that all pocket frames share is the track. Be sure to use a good-quality track that's strong enough for the doors. And for all pocket frames, unless the door(s) weigh more than 250 lbs., be sure the track is prepared with keyhole-shaped slots for all the track screws that are hidden within the wall. The track screws hidden inside the wall should never be tightened against the track, only the screws that are visible in the door opening. The screws in the door opening can then be easily removed and the track replaced, years after the wall is finished—a laudable improvement over cutting out a section of the drywall in order to replace a worn track.

Choosing between universal kits and built-up frames is a matter of personal preference. Universal kits are more sturdy because each vertical split stud is wrapped on three sides by steel. These kits can also be trimmed in the field to fit any door width, but they require more time to set and don't include split jambs or end jambs.

Built-up frames, on the other hand, are just that. They arrive on the job site built up, like a short, hollow section of 2x4 wall. Built-up frames are constructed with one vertical piece of steel channel on each side of the pocket mouth. Short horizontal pieces of 1x4 slip inside the vertical steel channel and are fastened to the channel and into the wood backing at the rear of the frame. Some built-up frames tend to be flimsy because the gauge of the steel channel isn't heavy enough. My advice is to see one before ordering it. The reason these frames are ubiquitous is because they can be set with ease and speed, without a table saw or many other common tools on the job site. And because they are shipped with all the necessary jamb components, when a carpenter finishes setting a built-up frame, it's entirely complete.

Universal kits Kits are available from most track manufacturers, such as Hagar, Grant, Stanely/Acme, Johnson, and Lawrence, and like the track, most kits are similar. The universal kit includes the header/track assembly with a jump-proof, U-shaped track, six pocket-door studs (1x2 lumber encased on three sides with steel), three floor plates (to attach each pair of studs to the floor), a rubber bumper that stops the door from

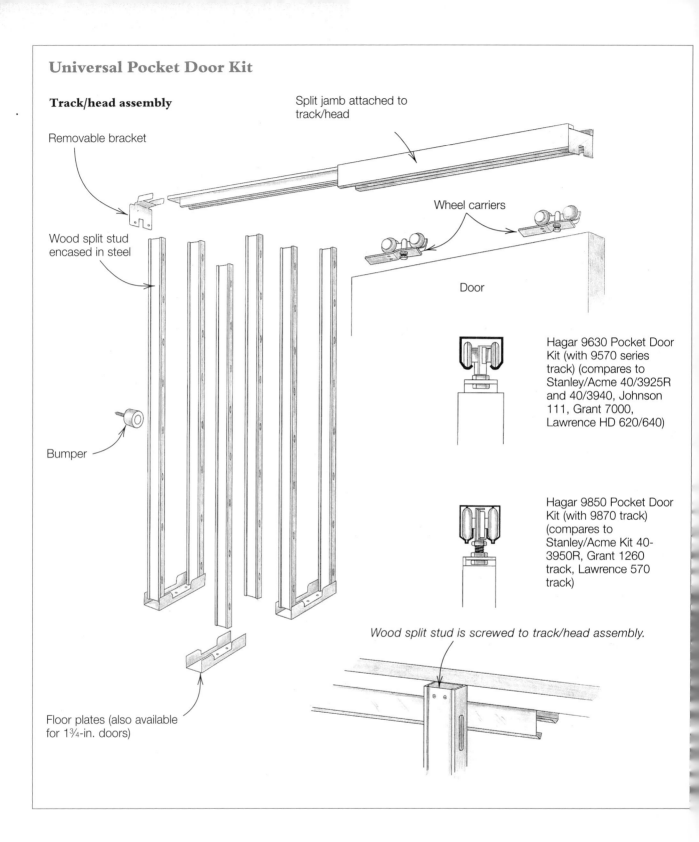

sliding too far into the pocket, and a pair of plastic guides that keep the door centered in the jamb throughout its travel.

Most of these kits are designed for use with 1⅜-in. doors, though accessory floor plates and shims are available so that the pocket opening will accommodate a 1¾-in. door (see the drawing on the facing page). A 1¾-in. door will only allow ⅛ in. of clearance between the door and the interior pocket frame. Even if the frame is set perfectly, a slight bow or warp will cause the door to rub on the inside of the frame. Fortunately, wider frames are available for 1¾-in. doors. Unfortunately, both universal kits and built-up frames made for 1¾-in. doors are wider than a 2x4 stud wall. To accommodate the additional width, one side of the continuing wall must be furred out ⅜ in. so that the framing and the pocket are flush. Usually this isn't a problem because pocket doors are often in bathrooms or other tight areas where walls are few and space is short.

Built-up frames Many local millwork shops, and even lumberyards, make built-up frames. In my area these frames are most common. Though manufactured pocket frames have taken a verbal beating since they were first introduced, most built-up frames used today are reinforced with steel to form somewhat rigid split walls. Unlike universal kits, which are packaged with many separate pieces, built-up frames are shipped in a strapped bundle that includes three pieces: the hollow-wall frame with the split jambs attached, the header/track assembly, and the full-width jamb leg for the strike side.

It's because these frames are so fast to install that they have become popular on production jobs. The header assembly slips into the top of the frame and screws are driven through the top of the head to secure the two pieces. The strike jamb is attached to the opposite end of the track/head, and then the unit is ready to install.

Installing manufactured frames Manufacturers of universal kits require slightly different rough openings, so refer to the installation instructions that accompany any universal kit. Like all openings, I like to set the trimmers plumb in both directions before doing anything else. Also, most universal track/heads must be cut to fit

After the split studs have been fastened to the track/head, a universal kit resembles a built-up frame and installation procedures are much the same. After the frame is in the opening, string the head. The track must be perfectly level and absolutely straight.

before installation. If the kit includes a split jamb attached to the track/head, both the track and the split jamb must be cut prior to installation.

Manufacturers' instructions vary and some suggest driving a few nails into the trimmers so that the track/head assembly can temporarily hang while the split studs are attached. But the easiest, quickest, and most accurate method for installing a universal kit is to set the track/head first. Begin by marking level lines at each side of the opening that represents the bottom of the jamb head (the face of the split jamb should be level with all other jambs in order to align the casing). If the split jamb is not attached to the track/head, then allow for the thickness of the jamb head and make another mark on each trimmer at the additional height of the track (projected above the jamb head). Before installing the track/head, be sure to slide the wheels into the track—sometimes they can't be installed later. Screw the end brackets into each trimmer securely. At this point, the only requirement is that the two ends are level.

If the framer's snap lines aren't still visible on the floor, then snap new lines across the opening, from trimmer to trimmer, on each side of the wall. Next, slip the split studs into the floor plates, then position each plate on the floor beneath the track/head. The pair of studs at the mouth of the pocket is the most critical, so leave them until last. First install the rear pair and any intermediate studs. Screw each stud to the track/head, but don't

Pocket Door Guides

There are several types of guides used on pocket doors, as shown in the drawing at right. The most common are inexpensive and easy-to-install U-shaped plastic guides that wrap around the bottom of the door and nail to the face of the door stop or split jamb. These guides work acceptably well if they're installed a little loose, although they can be an eyesore. Unfortunately, if a little paint or dirt gets on the plastic guides, a small mark will begin to appear on the bottom edge of most doors. Over time the mark can become a small groove. Other plastic guides are also available that are designed to contact only the very bottom edge of the door, and these are preferable, though they too can be an eyesore.

It's tempting to avoid slotting the bottom of a door and instead use a plastic surface-mounted guide, but as usual the extra work is worth the effort, and for single pocket doors or pairs of pocket doors, nothing works as well as a slotted door and a floor guide. It isn't necessary to attach a long guide to the floor to control the travel of a pocket door. In fact, using a long, angled guide can foul the operation of the door. A short guide, about 3 in. long, near the mouth of the pocket is more than ample. The guide must project slightly beyond and within the opening to control the door in both directions of travel.

Simple floor guides, however, don't work for multi-door, bipass pocket openings, where a door must travel completely outside the mouth of the pocket frame. For these openings, a floor track and bearing guide are necessary. This track can be let into any hard-surface finished flooring, or it can be set flush with carpeting (see the drawing on p. 181).

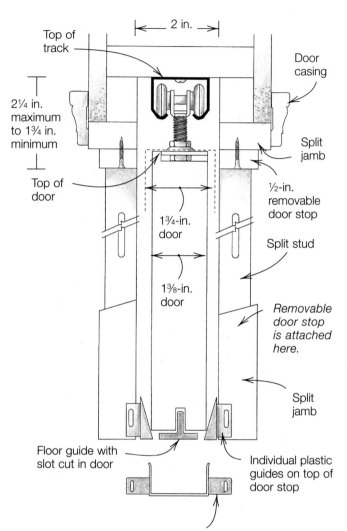

Pocket framing and guides

fasten the floor plates yet. To straighten the track/head, shims may have to be inserted between the floor plates and the subfloor.

I believe a pocket door should resemble a swinging door as much as possible. After all, pockets are only meant to hide swinging doors. Therefore, I install door stop on the head of the jamb and on the face of the split jamb legs (see the drawing on the facing page). No door stop should be attached to the strike jamb because it will interfere with the function of the door, and besides, two pieces of door stop, like railroad tracks running up the strike jamb, look terrible. Also, door stop allows permanent fastening of the split jambs and the casing—only the door stop needs to be removed if the door has to be adjusted or changed! When using door stop, locate the split studs at the mouth of the frame so that the finished opening will be ⅛ in. wider than the door. In other words, the space between the trimmer and the first pair of split studs must allow enough room for the split jamb, the strike jamb, and the width of the door, plus ⅛ in. for clearance. Because the trimmers have already been set plumb, I install the strike jamb first (not supplied with a universal kit), then I locate the split studs back from the face of the strike jamb the width of the door plus ⅞ in. (the thickness of the split jambs plus that ⅛ in. necessary for the door to clear the jamb).

Once all the pairs of split studs are fastened to the track/head, a universal kit is much like a built-up frame. The primary difference is that built-up frames are connected across the bottom of the pocket so only one floor plate needs to be fastened to the subfloor. Plumb the split studs exactly—there should be no need for shims between the studs and the split jamb—and align the studs in the center of the wall, then mark that location. Before fastening the floor plate, check that the track head is absolutely straight (see the photo on p. 189). Use thin nylon line, or even fishing line, because typical mason's line is too thick for the accuracy that's required in finish work. Shims can be inserted beneath the floor plate(s) to lift and straighten the track/head (see the photo above).

Installing a Pocket Door

Installing the door can wait until after drywall, but to be certain the opening is perfect, and in order to make fine adjustments in the fit, it's good practice to install

Don't fasten the track/head to the header, except near each trimmer, just in case the header settles. Instead, raise or lower the track by shimming the frame from the floor. Built-up frames are easy to align and quick to set because they come in one piece and only require a single fastener driven into the floor.

the door before the walls are sealed up. First, fasten the wheel carrier brackets to the top of the door. It's best if the screws are driven into the door's top rail and not in the end grain of the stile, but in some cases the door stiles are wide and the door is narrow and if the wheels are too close to each other the door will rock, especially a narrow door. If the screws have to purchase in end grain, then drill a small diameter pilot hole and fill the hole with glue before driving in the screw. I've never had a screw pull out of a stile prepared this way.

It's easy to hang a pocket door on its track if the hardware isn't cheap. And it's easy to judge track: If the door has to be tilted so that the wheels can be hooked over the track, use a different type of track. Quality pocket/sliding door hardware uses jump-proof wheels. The wheels run inside the channel of the track or are trapped between the I-beam of the track, so that they can never come off the track (that's why the wheels often must be installed before the track/head or built-up frame is placed in the opening). One man can install a pocket door with quality track and wheels, even if the door is really big.

Once the door is standing against the opening, lift the back end, near the mouth of the pocket, and pass the wheel-carrier stud through the slot in the wheel carrier plate. While lifting the opposite end, the door will roll

Bipass Pocket Doors and Pairs

Pocket door framing and finish jambs

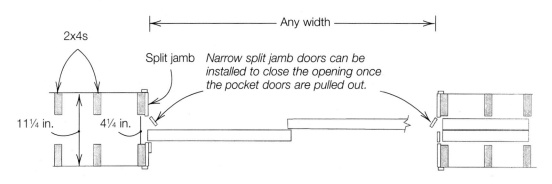

Track, guides, and header assembly

Track interchanges with Grant 1210/1230, Hagar 9100.

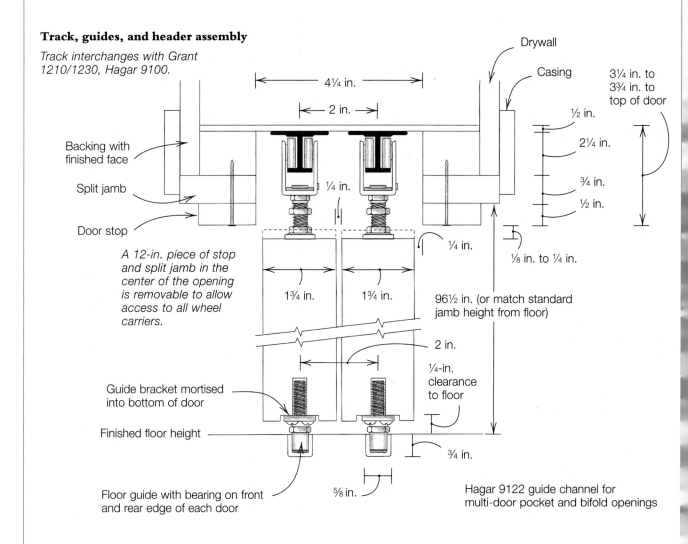

inside the pocket, which makes it a little easier to engage the front wheel-carrier stud in the plate. Adjust the wheels so that the door meets the strike jamb flush. When the door is drawn back inside the pocket, the lock stile should be even and flush with the split jamb. If it's not, then one of three things is wrong: the jambs aren't plumb, the head isn't level, or the track/head isn't straight. Most often, I've found that the problem is the track/head, which settles under the weight of a door, especially if the floor plate(s) isn't shimmed properly. Surprisingly, that one simple step is most critical.

I like to adjust the door so that it's flush with the door stop at the mouth of the pocket, which sometimes means trimming a little off the face of the rubber bumper stop or mounting the stop on a spacer attached to the back of the pocket frame. Another technique is to make an adjustable bumper by threading a furniture glide into a T-nut mounted on the rear of the door or jamb.

> ### Miscellaneous Hardware
>
> **There's a large assortment of gadgets manufactured for sliding and bifold doors. There are several stop bumpers available for pairs of sliding doors that screw against the upper track and limit the travel of each door, so that a pair of doors can be stopped right at the center of an opening or one leaf in a multiple-door opening can be stopped in its perfect position. Of course many suppliers stock a variety of edge pulls, flush pulls, and privacy latches, but key locks are also available for pocket, bifold, and bipass doors (Hagar 9262, 9264, 9268). If you can't find the hardware you need for a special application, contact a track manufacturer and ask for a catalogue.**

Site-Built Pocket Door Frames

Manufactured frames can be ordered to suit almost every need, but pocket frames still have to be built in the field for unusually heavy doors or for pairs of bipass pocket doors (see the drawing on the facing page). With the track and accessories available from different manufacturers, custom pocket-door walls can be designed by anyone willing to work out the details, and almost any size opening is possible. The drawing is an example of a 15-ft.-wide opening with 8-ft.-tall doors. Exact drawings and accurate details are essential components of a successful job.

Installing the track and head jamb Like all pocket-door frames, the installation begins at the header. The height of the finished jamb (because the head casing should align throughout a home) and the required depth of the track pocket determine the height of the header. Therefore a detailed drawing, even a crude one, is a prerequisite to setting the header. As shown in the drawing on the facing page, the finished jamb (not the door stop), must be 96½ in. from the floor, to be level with the other jambs in the home. The pocket requirement averages 3½ in., and because that includes ½-in. door stop, the header must be 99½ in. from the floor.

Headers are rarely straight, so always include additional space for shims: ¼ in. is ample for a short span, but for a wide opening that requires a big, long header, add ½ in. to the height of the header to accommodate inconsistency and deflection in the timber. It's easy to shim the track down; it's not so easy to raise the header later. As with long, site-built bifold openings, the track must be set perfectly level and straight or the doors won't hang plumb and parallel to the split jambs. Once the track is installed, filling in the studs is identical to setting a universal pocket door kit.

Closing a wide pocket Pairs of bipass pocket doors require wide pocket mouths, which can be unsightly. Sheeting can be installed on the inside of the pockets walls and even painted before the walls are raised in sections, but the best solution is to make narrow hatch doors to cover the mouths of the pockets. The hatch doors should be built in pairs, so that one leaf can be closed against the rear door when all the doors are in the openings and both hatch doors can be closed when all the doors are in the pockets. Use small piano hinges on the hatch doors, so they won't warp, or a series of small Soss hinges, in sizes #14 or #30, so the hinges will be completely invisible.

Chapter 8
WEATHER-PROOFING FOR EXTERIOR DOORS

SILL COVERS

THRESHOLDS

DOOR SHOES

PERMANENT INSTALLATION

WEATHERSTRIPPING THE JAMB

Weather-proofing an exterior doorway correctly is a critical detail in construction. Because weather infiltration causes extensive and often hidden damage, contractors and homeowners can avoid enormous losses by using and installing the right weather-proofing products. At first glance, selecting these materials might seem as difficult as choosing a new entry door or lock, but the subject of weather-proofing is more black and white, not only because there are fewer color choices, but because most weather-proofing products are meant for specific environments. Like many construction procedures, it's best to start at the bottom, which is why I'll begin with sill covers, then move on to thresholds, door shoes, and finally, weatherstripping. As I discuss the products that are available, I'll also describe the step-by-step techniques I use to install weather-proofing. Like all aspects of carpentry, each step must be completed in the proper order to successfully weather-proof a doorway—and the first step is always the sill cover.

SILL COVERS

Though sill covers are considered decorative trim by many clients, for carpenters these aluminum nosings are life-savers. Not only do they cover the rough edge of a concrete slab or the exposed grain of a wood floor, but they're also the flashing for the threshold. They're also the perfect cure for elevation problems—they provide the means to raise a threshold when finished flooring ends up too high. Sill covers are manufactured in sever-

Water-Return Threshold

Both the drain pan and sill cover of a water-return threshold act as flashing, capturing any water that makes its way under the door and directing it back outside.

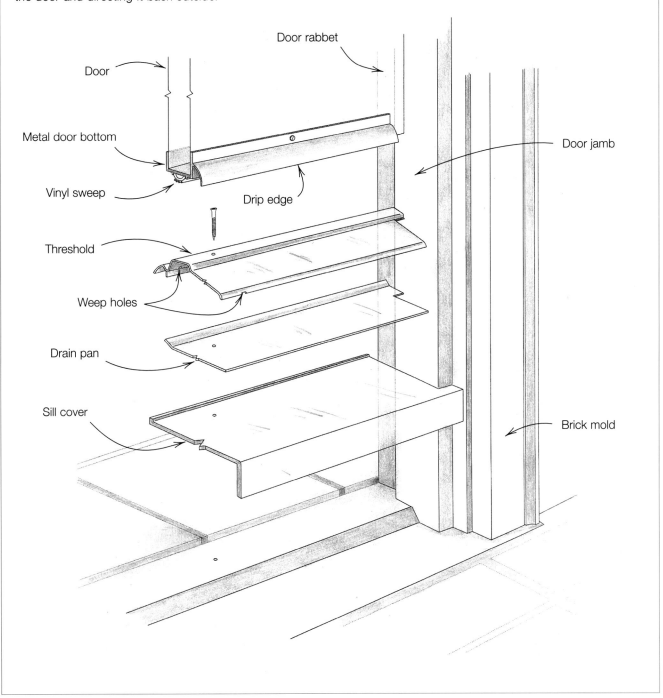

WEATHER-PROOFING FOR EXTERIOR DOORS

Cut the sill cover to fit between the exterior trim, then scribe each end to notch around the screen rabbet and over the door stop.

Tip the sill cover into the opening, one end at a time, to scribe the depth of each notch. Adjust the sill cover so that the depth of the notch is equal at both ends of the jamb, or the aluminum won't be parallel to the opening.

Use small shims and a short level to slope a sill cover. If necessary, scribe the nosing to achieve sufficient slope. Masking tape makes the scribe line easier to see.

al sizes, from 2¼ in. wide for standard jambs in 2x4 walls to 4½ in. wide for wide jambs. Generally the vertical nose on a sill cover is between 1½ in. and 1¾ in. tall.

Fitting and Cutting

Sill covers should always be notched tightly against the jamb, sealed in a bed of silicone, and tipped slightly toward the exterior so that water drains away from the door opening. For most weather-proofing products, it's best not to use a tape measure, but rather rely on scribing to ensure a tight fit. Hold the sill cover up to the outside of the opening, with one end butted against the exterior molding. Mark the sill cover on the opposite side of the door, at the opposing molding. Carry both marks across the top of the sill cover with a sharp pencil. If the pencil line gets lost in the color of the aluminum or in the bright finish, scratch the surface by drawing a utility knife alongside a square. Scratches show up easily in anodized aluminum, which is good, but it also means that cutting must be done carefully (see the sidebar on the facing page).

Once the sill cover is cut off square it should fit between the exterior trim (see the top and center photos at left). If the exterior trim has been installed flush with the face of the jamb (to allow room for a screen door the trim is often flush with the small screen rabbet), then the sill cover only has to be notched for the screen rabbet. If the exterior trim has been installed back from the face of the jamb, leaving a reveal line (which is necessary on a swing-out door to allow room for the hinge barrels), then cut two notches—one for the reveal and one for the door rabbet.

After cutting the notches, set the sill cover in place and prepare to trim the front nosing. On some openings this step isn't necessary, but if there's a concrete porch or wooden step just beneath the sill of the door, then the sill nosing has to be scribed to the porch, too. The nosing must fit tight against the subsurface to prevent water infiltration and to create a secure footing for the threshold that will be installed on top. As well, the sill cover must slope so that the water will drain outside, not inside (see the bottom photo at left).

Cutting Aluminum Weather-Proofing Products

Cutting aluminum is not my idea of fun, so I like to do it as quickly and as safely as possible, and also eliminate any risk of scratching the product. Most people use hacksaws to cut weather-proofing, and they work well, except that they're slow and tiring, especially when working on more than one door. It's also easy to slip with a hack saw and make a mistake. Some professional weatherstrippers use portable table saws with aluminum-cutting blades, which work well. Occasionally, I'll even use my power miter saw with a carbide blade.

But the best all-around tool I've found for cutting aluminum is a 4⅜-in. Makita 4200N trim saw, which cuts at a fast 11,000 rpm, and a fine-toothed, combination metal blade, also by Makita (model 792334-2). I've used carbide-tipped blades to cut aluminum, but they are expensive, especially when the teeth begin to break. Unfortunately Makita stopped making the 4⅜-in. saw and now markets a slower-turning saw with a 5½-in. blade. Jepson manufactures a similar 4⅜-in. saw, though I like my Makita better because it runs smoother, with less vibration and slightly less noise. I've heard that Makita will be making its 4200N again soon.

Noise is a problem associated with cutting aluminum, so always wear ear protection (you should when using any power tools). Eye protection is also necessary to guard against small bits of flying metal and occasional chipped saw teeth. In my work box I carry plastic goggles wrapped in a sock to prevent scratches. Years ago a bungee-cord accident took most of the vision in my left eye, so I'm careful with my right one.

Whether you use a hack saw or a trim saw, support the work piece on a stable surface or bench, one that you can stand at comfortably and see the work as you're cutting. And feed the blade into the material slowly. With a power saw, be sure the rpms are up before cutting the metal. Cut through the material slowly, allowing the saw to run at maximum speed, all the way through the cut. Don't turn the saw off until the waste has fallen off. If you use a chop saw, stand on one side of the blade and hold the measured work piece securely (chop saws have a reputation for kicking small fall-offs and sending them flying through the air).

Always wear ear and eye protection when using power tools, especially when cutting aluminum. Don't push the saw, but rather allow the blade to cut smoothly. Let the waste fall off before stopping the saw.

Scribing the sill cover so that it both slopes and fits precisely can be done in one step. Shim and tip the inside of the sill cover, then check the slope with a torpedo level. Too much slope will cause the threshold to be raised on the inside, which can interfere with the operation of the door shoe; too little slope will allow standing water. About one-eighth of a bubble is just right. Spread a pair of scribes to match the distance between the floor and the bottom of the sill cover on the inside of the jamb. Squeeze the scribes closed ⅛ in. to leave room for shims on the inside, then scribe a line across the face of the sill cover nosing. If the cut is made right to the scribe line, the sill cover will fit perfectly.

THRESHOLDS

The threshold goes in next, on top of the sill cover or directly on top of a wood sill. Two types of thresholds are predominantly used today, water-return thresholds and oak-top thresholds, and each has advantages and disadvantages. The ground isn't the only thing that gets wet when it rains hard—doors get wet, too, and as a door swings open water is often swept inside by the door shoe sweeping over the threshold or just by it dripping off the door. Water-return thresholds are designed to collect the water that's swept inside and carry it back outside.

Instead of having water-return dams and drain pans, oak-top thresholds are wider than plain aluminum thresholds and are available in sizes wide enough to reach the exterior wall of a sill, which guarantees that water will always flow out of the door opening (see the photo below). These relatively new products are also popular because the mixture of oak and aluminum is both attractive on the inside of a home and practical on the outside.

Oak-top thresholds require more complicated blind notching because the wide threshold covers the sill nosing almost completely and must be notched around the rabbeted jamb, and sometimes around the screen rabbet, too.

Water-return thresholds are made in three different designs for different purposes. Low-rug water-return thresholds are best for vinyl floors or doors that swing in over hard-surface floor covering, like stone or hardwood, where the door only needs to clear a low throw rug. Low-rug water-return thresholds, like the one shown in the drawing on p. 195, come with a separate drain pain that is installed beneath the threshold and directs water outside. High-rug thresholds are taller and are meant for doors that swing in over carpet. The high-rug threshold in the top left photo on the facing page is 1 5/8 in. tall and contains an integral drain pan rather than a separate drain pan. An intermediate, 1 1/16-in.-tall, high-rug water-return threshold is also available with a separate drain pan.

Interlock thresholds are also water-return thresholds, but they require a hook on the bottom of the door, rather than a door shoe, as shown in the top right photo on the facing page. Hooks for interlock thresholds (also called door hooks or hook strips) are installed in a rabbet that's cut into the bottom of the door. The rabbet must be cut at exactly the right height or the hook strip will not engage the top of the threshold properly. In many climates, where humidity fluctuates dramatically throughout the year and doors swell and contract considerably, interlock thresholds can cause occasional problems because there is no room for adjustment. Interlock thresholds are more expensive to install than water-return thresholds because the bottom of the door has to be rabbeted and cut precisely, but there are locations when interlock thresholds are the best answer. On challenging job sites, where homes are exposed to turbulent weather (like high unprotected hills or near the ocean) and on homes where design restrictions prohibit high thresholds, an interlock threshold is the only choice. Interlock thresholds have been in use long before I was born and are proven to work.

Many warehouse outlets primarily sell rolled residential saddles, but they have no water-return dams, no weep holes, and no drain pans, and they often leak (see the bottom photo on the facing page). I only recommend these simple thresholds for openings that are completely protected from the weather, like interior doors that lead between the garage and the house.

This high-rug water-return threshold has an integral drain pan. The door is rabbeted for an L-shaped door bottom so that no metal is visible at the bottom of the door on the inside of the home. The jamb weatherstripping is kerf-in thermoset foam manufactured by Q-Lon.

Like water-return thresholds, interlock thresholds have a water-return damn, weep holes, and separate drain pans. But the seal between the door and the threshold is accomplished by a hook, rabbeted into the bottom of the door, which interlocks the top of the threshold. The jamb weatherstripping is silicone bead.

Cutting and Fitting Thresholds

Most aluminum thresholds are installed similarly, regardless of the design. Even interlock thresholds are installed just like other water-return models. Scribing and cutting the bottom of the door is the only difference, but I'll get to that later. First I'll go through the steps for installing a threshold.

Try to fit the threshold the same way as the sill cover, by scribing each end. If the threshold is too long and won't fit against the opening, then take one measurement across the door rabbets. After making the first cut for overall length, slide the threshold into the door rabbets and mark each end for a notch that will clear the face of the door stop, as shown in the top photo on p. 200.

Rolled residential aluminum saddles should only be installed on openings that are protected from the weather, like a house-to-garage door. This door shoe is a common 1¾-in. full U shape that wraps completely around the bottom of the door. Adjustable rigid weatherstripping is screwed to the jamb.

WEATHER-PROOFING FOR EXTERIOR DOORS

Scribe the threshold to fit the door jamb as you did for the sill cover—by marking both ends for the leg of a notch that will fit around the rabbeted door stop.

Tip the threshold against the door stop to mark the shoulder of the notch. Before marking the shoulder, be certain the edge of the threshold will cover the flooring. If pencil lines aren't clearly visible against the anodized finish, use a knife to score the lines.

Normally, the threshold sits directly beneath the door, but the position of the threshold can also be determined by the finished flooring. The edge of the flooring should be covered by the threshold so that no additional trim is required. It's not unusual to move a threshold ¼ in. inside a home so that it will cover the raw edge of a tile or vinyl floor that's been cut a little too short of the jamb. Once the position of the threshold is determined, tip it into the opening so that the end is against the door stop and the interior edge projects inside the house far enough to reach the finished flooring. The threshold will be too long to lie flat in this position, but only one end at a time needs to be marked, and only for the shoulder of the notch, as shown in the bottom photo at left. Repeat the same procedure for the opposite end.

Cutting and Locating the Drain Pan

Once the threshold is cut, temporarily set it in place on top of the sill cover, then lightly mark the spot where the inside edge of the threshold rests on the finished flooring. This mark helps determine the position of the drain pan. (If the sill cover projects inside the room beyond this mark, trim it back so that it stops beneath and just before the foot of the threshold.) It's important to locate the drain pan carefully so that it catches water seeping through the weep holes in the threshold dam but at the same time remains hidden from view, as shown in the photo on the facing page. After marking the floor at the inside edge of the threshold, remove the threshold and use it as a straightedge to mark a heavier line ⅛ in. closer to the jamb. Now cut and notch the drain pan to fit tightly against the rabbeted jamb, with the back of the pan flush with the second heavy pencil line. The drain pan cuts easily with tin snips or sharp scissors.

Place the drain pan on the sill cover, then set the threshold on top of the drain pan. Run a ³⁄₁₆-in. bit through the holes in the threshold and then through the drain pan and sill cover. For a wood subfloor, drill ⅛-in. pilot holes for the mounting screws; for a concrete slab, run a ¼-in. carbide bit through all three pieces. That's the surest method I know of to get the anchors in the right spots. But don't fasten the threshold yet or the door won't clear and close. Once the door has been scribed for the threshold and the door shoe, the sill cover, drain pan, and threshold can be installed permanently, but first you need to determine how much to cut off the door to allow enough room.

The back of the drain pan, which is bent at about a 45 degree angle to stop water from flowing inside the house, should be positioned beneath and just before the foot of the threshold. The point of the pencil rests on the heavy line I marked at the back of the threshold.

DOOR SHOES

There are several popular door shoes, and each has a particular use and requires a slightly different installation technique. An oak-top threshold or a water-return threshold requires an aluminum door shoe with a vinyl sweep to seal the gap between the bottom of the door and the top of the threshold. An interlock threshold uses a door hook to seal the space between the door and the top of the threshold, as shown in the top right photo on p. 199.

Vinyl-Sweep Shoes

The two most common vinyl-sweep door shoes are either U shaped (see the bottom photo on p. 199) or L shaped (see the top left photo on p. 199). Concealed vinyl-sweep shoes are also available, but they are used less frequently because they aren't easily adjustable, like U-shaped and L-shaped shoes, and because they require a more difficult installation procedure (see the sidebar at right).

Dadoes for Concealed Door Shoes

Occasionally the design of a home necessitates a clean, aluminum-free look. On some projects, I'm not even allowed to install drips on exterior doors—an aesthetic choice with which my pragmatic, carpenter-side disagrees. However, there are openings where drips can't be used, like two exterior bifold doors (see Chapter 7 for more on this). For these openings, concealed vinyl-sweep door shoes are the best choice.

Concealed door shoes are not easily adjusted; the door has to be removed and the shoe lowered either by the use of shims or adjustment screws, then the door rehung. To eliminate time-consuming adjustments, scribe and cut the door carefully and precisely. Start by scribing the bottom of the door ¼ in. from the top of the threshold.

After cutting the door square at the scribe line, cut a ¼-in.-deep by 1¼-in.-wide dado in the bottom of the door for the shoe. I use a router with a ¾-in. mortising bit to cut the dado, and I make a temporary modification to my standard flush-bolt fence for this specific job. The homemade flush-bolt fence on my router is set to center a ¾-in. bit in the middle of a 1¾-in. door. I cut the 1¼-in.-wide dado by adding a thin, ¼-in. strip of hardwood to the fence and running the router on both sides of the door.

Before installing the shoe and rehanging the door, seal the bottom thoroughly with two coats of exterior primer.

WEATHER-PROOFING FOR EXTERIOR DOORS

For vinyl-sweep shoes, mark the jamb ½ in. above the threshold.

Spread a pair of scribes from the top of the sill cover to the mark on the jamb, swing the door closed, then scribe a line across the bottom of the door. Use masking tape on the door if the line is hard to see.

U-shaped shoes are available for 1¾-in. and 1⅜-in. doors and wrap completely around the bottom of the door, providing maximum protection from the weather. L-shaped shoes can be used on doors of any thickness but are not visible on the interior side of the door.

U-shaped and L-shaped shoes both require ½ in. of clearance between the bottom of the door and the top of the threshold, so the first installation step is to mark both jamb legs ½ in. above the threshold, then remove everything but the sill cover (see the photos above). The sill cover provides a smooth surface from which to scribe the door. It's simple to cut the door right to the scribe line. Seal the bottom of the door with a minimum of two coats of exterior primer. I use a product like Kilz, because it's a fast-drying exterior primer and I'm able to hang the door back up quickly.

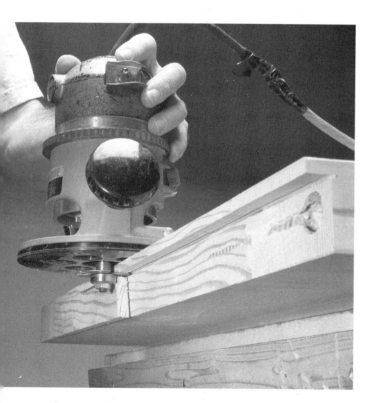

A router with a slot cutter makes rabbeting the bottom of a door easy work, and the rabbet is always straight and clean (part of the waste has been removed for clarity).

Installing L-Shaped Door Shoes

Doors for L-shaped shoes are often cut off square, just like doors for U-shaped shoes. Because an L-shaped shoe is only 1¼ in. wide, it isn't visible on the interior side of the door unless you stand back a good distance, and even then you have to bend down a little to see the door shoe. But the black, shadow-line space between the bottom of the door and the threshold is visible—and it doesn't look good. A door cut off square for an L-shaped shoe always looks like it has been cut too high, like pants legs that are too short. Rabbeting a door for an L-shaped shoe solves that problem. Once rabbeted, a door will reach almost to the threshold on the interior side, like a door cut for an interlock hook, without the interruption of aluminum. This clean look is often requested by designers, architects, and homeowners, and it puts an L-shaped bottom to good use. The same scribe line marked for a U-shaped shoe works for an L-shaped shoe, too.

First, measure down from the original scribe line ⅜ in. and cut the door off square at the new, lower mark, so that the door will clear the top of the threshold by ⅛ in. Then set the depth of the circular saw to 1⅜ in. and follow the upper, higher scribe line to cut the shoulder of the rabbet. I use a router fitted with a slot cutter to cut the short leg of the rabbet, as shown in the bottom photo on the facing page. A circular saw can be suspended over the edge of the door with a fence attachment, but the setup is awkward and risky. The rabbet can also be cut by making repetitive passes with a circular saw, though that's the slowest method and doesn't leave a very clean job. Be sure to seal the bottom of the door thoroughly, especially because an L-shaped shoe doesn't wrap around the door and a rabbet is even more prone to wicking moisture from the air.

Installing Interlock Hooks

Carpenters are often alarmed by interlock thresholds with door hooks because the tolerances are very close—the door must be cut at exactly the right height and the bottom of the door rabbeted at precisely the right depth or the hook and threshold won't work. But scribing and rabbeting for interlock hooks isn't really any more difficult than scribing and cutting a door for a standard shoe.

For interlock thresholds, a regular pencil laid on top of the threshold leaves a perfect mark on the jamb, ⅛ in. above the threshold—exactly where the shoulder of the rabbet must be cut.

Before scribing the bottom of a door for an interlock threshold, be sure the threshold is sitting flat against the floor. If there's any doubt, screw the threshold down temporarily, then use a regular pencil to mark the sides of the jamb, as shown in the photo above. Remove the threshold so that the door will close, but leave the sill cover in place. As you would do with any other door, spread a pair of scribes from the sill cover to the mark on the jamb, then close the door and scribe a line for the shoulder of the rabbet.

Cutting the rabbet is similar to rabbeting for an L-shaped vinyl shoe. Mark the door ⅜ in. below the scribe line, and cut the door off square at the lower line. Then rabbet the door at the high line. Seal the bottom of the rabbet completely before installing the door hook (while the paint is drying the threshold can be installed permanently). I get impatient waiting for two or three coats of primer to dry. I usually end up with paint on my fingers because I install the hook while the last coat is still wet.

WEATHER-PROOFING FOR EXTERIOR DOORS

board and plywood underlayment typically used beneath vinyl flooring also react poorly to moisture. Overall, because most water damage occurs near the sill of a doorway, it's makes good sense to spend a few extra minutes at every door.

Installing the Sill Cover

I use a full tube of silicone sealant on most doorways, which means I seal them well. To ensure a good seal, clean the subfloor thoroughly, then spread a generous bead of sealant beneath the Moist-Stop, plus another bead between the Moist-Stop and the sill cover. Prevent water from penetrating around the ends of the sill cover by running a bead of sealant alongside each jamb leg, too. Run another bead on top of the sill cover, about 1 in. behind the front of the drain pan. Silicone sealant is a good adhesive, too, and helps secure the drain pan. Run one more bead along each jamb leg, then press the drain pan down. After the drain pan is set, seal the ends against the jamb, all the way to the back of the bend in the drain pan, so that the dam is continuous (see the top photo on the facing page). Finally, fill each screw hole in the drain pan with silicone and set the threshold on top, then secure it in place with the mounting screws.

Installing the Door Shoe

A vinyl-sweep door shoe can't be installed until after the jamb weatherstripping has been applied because the shoe must be notched to clear the weatherstripping (I'll cover the subject of weatherstripping next, but for the sake of continuity I'll finish with the door shoe now). Once the threshold and weatherstripping are installed, the door shoe can be scribed and trimmed (see the bottom photos on the facing page). Before installing a vinyl-sweep shoe, crimp the aluminum track around the vinyl so the vinyl won't slide. Be sure to pull the vinyl sweep out each end of the door shoe first, so that the vinyl will project beyond the edge of the door and seal the hinge and strike gap all the way to the jamb. This important and simple step is often overlooked, although it means the difference between a thorough seal and seeing daylight at each edge of the door.

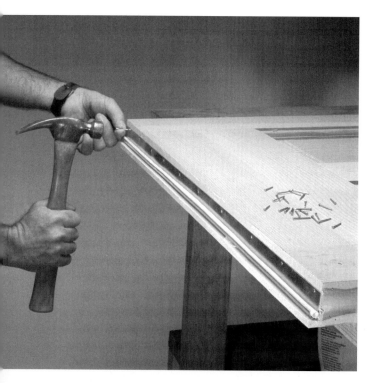

The hook should be cut to the width of the door and if there's a flush bolt, it must be notched, too. Work from one end of the door to the opposite end, driving a nail every 2 in. to 3 in. Be sure the hook is up against the shoulder of the rabbet before driving each nail, or the door will bind on the threshold. Silicone bead can be attached to the inside of the hook to improve the seal.

On openings that are particularly prone to high winds and wind-blown rain, I install a piece of narrow-diameter silicone bead weatherstripping inside the hook strip. I use a little silicone caulking to secure the weatherstripping bead, which compresses easily, though it can make the door a little harder to close.

PERMANENT INSTALLATION

Once the door is cut for the threshold and shoe, the sill cover and threshold can be installed permanently. But before I install the sill cover I like to lay down a waterproof membrane of Moist-Stop (see Chapter 4). Cold, icy weather and frost can transfer through aluminum, and a waterproof barrier is cheap insurance against expensive water damage on a hardwood floor. Particle-

Pairs of doors often have flush bolts on the inactive leaf, so the inactive door shoe must be drilled for the flush bolt. After trimming the shoe to fit the width of the door, hold the shoe up against the bottom of the door and trace the location of the flush bolt on the inside of the shoe. Be sure the aluminum is crimped on the vinyl before drilling the hole, so that the drill bit can penetrate both the aluminum and the vinyl at once. Drill a small pilot hole first, then finish the hole with a ½-in. bit. Ream the hole a little larger, so that the flush bolt won't bind. Trim the vinyl fingers on the sweep so they angle back from the hole, or they may obstruct the flush bolt from entering the hole.

After cutting the door shoe, slip it on again and swing the door shut to make sure everything fits. The door shoe shouldn't be too long and squeeze or rub against the weatherstripping, but it should come close. Press the shoe down against the threshold, but not too hard. The

The sill cover, drain pan, and threshold are like a sandwich, with silicone spread generously between each layer.

Scribe a line on the face of the door at the front of the weatherstripping, where the door shoe will have to be notched to clear the weatherstripping.

Hold the shoe flush with the back of the door and trace a line across the inside of the shoe at the front edge of the door. The door shoe is cut to match the bevel of the door. Also, mark the top of the door shoe at each end of the door for the weatherstripping notches.

WEATHER-PROOFING FOR EXTERIOR DOORS

vinyl sweep needs to contact the threshold, but shouldn't be forced against it. Otherwise, the sweep will compress over time, and the seal will be lost. Also keep this important thought in mind: A small amount of water will invariably find its way between the door shoe and the face of the door. When I fasten the shoe to the door, I set the front end first, so that it just contacts the threshold and seals the opening. Then I tap the rear end of the shoe, at the hinge leg of the jamb, a little tighter against the threshold—just enough to create a slight bit of fall on the inside of the shoe so that water can trickle out the back of the shoe and drip into the water-return dam, even when the door is open (see the left photo below). At the lock stile, use silicone to seal between the shoe and the bottom edge of the door so that water will never drip off the leading edge of the door. Never seal the back of the shoe, though, or water will be trapped between the door and the shoe, eventually ruining the door.

Auto-Bottoms

In many parts of the Southwest, where the weather is agreeable most of the year, architects occasionally design homes with exterior doorways that have no thresholds or visible weather seal on the bottom of the doors (see the right photo below). Openings without thresholds are often found on homes that have hard-surface finished floors—like stone, polished concrete, or ceramic tile—that extend and flow from the inside of the house to the outside, seamlessly transforming interior flooring to exterior entryways, patios, or pathways. Automatic door bottoms, also called auto-bottoms, are the best means of achieving a weather seal in these openings. Though they are not 100% effective and can't completely stop wind-driven rain (especially on a pair of doors fitted with flush bolts), auto-bottoms work well in openings that are protected by a substantial overhang and in homes that aren't blasted by weather.

Installed correctly, a water-return threshold really does its job. Any water dripping from the door will drop into the dam, weep into the drain pan and out the front of the threshold, and then flow down the sill cover and outside.

Automatic door bottoms can be difficult to install but they're almost invisible and work well in openings where a threshold and door shoe would distract from the design of a continuous floor or interfere with traffic. On the left is the backing screws; on the right is the auto-bottom and plunger screw head.

Surface-mounted auto-bottoms are available that mount directly to the face of the door, but they're an eyesore and popular mostly in commercial applications. In residential door openings, I always install mortise-type auto-bottoms, which aren't any more difficult to install than flush bolts, if the right hardware and the right tools are used.

Several types of auto-bottoms are manufactured. I like Pemko's 411 model because it fits well in a 1¾-in. door. The 411 only requires a ⅝-in.-wide mortise that's 1½ in. deep (see the drawing below). Many manufacturers offer similar hardware. Most auto-bottoms will seal up to a ¾-in. gap, but I try to scribe these doors within ¼ in. of the floor, so that the neoprene seal doesn't have to travel so far to reach the floor. The less hardware that's seen, the better the look.

Mortising for an Auto-Bottom

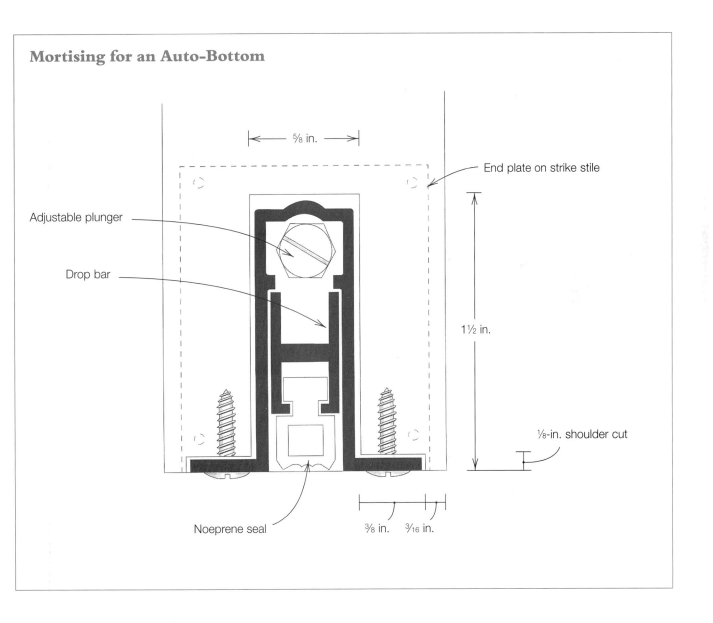

I use my plunge router and standard flush-bolt fence to mortise for auto-bottoms, and swap the ¾-in. flush-bolt bit for a ⅝-in. bit. The homemade fence on my router centers any bit on a 1¾-in. door (see Chapter 6). I cut about 3⁄16 in. with each pass of the router. A circular saw can be used to cut the mortise, but it changes the job from a clean and simple operation to a slow, hard task. At least use a standard router with a ⅝-in. bit to cut most of the mortise, then finish up with a circular saw because a standard router can't cut deep enough.

After finishing the main mortise, adjust the router fence to make the two shoulder cuts for the screw flanges. Because my flush-bolt fence isn't adjustable, I add a strip of hardwood to the fence to move the ⅝-in. bit closer to the edge of the door. The shoulder cuts are only ⅛ in. deep, an easy job for any router, and really allows the auto-bottom to be hidden beneath the door.

For the hardware to work properly, the mortise should be open at both ends of the door. Though I have cut mortises that are blind at the strike side of the door, to avoid using the metal end plate supplied with the auto-bottom, I don't recommend the practice because the neoprene gasket must be cut short of the extreme edge of the door for the drop bar to retract completely, which compromises the seal. No matter how the door is dadoed, always sand the mortise before installing the hardware.

To install the auto-bottom, first cut it to length. Up to 2 in. can be cut from the end opposite the plunger, so the device can be customized to fit any door. Use a hacksaw or carbide-tipped blade in a power miter saw to cut through both the heavy aluminum body and the drop bar. When mounting the hardware, drill pilot holes for every screw—the shoulders of the door flanking the mortise are a little more than ½ in. wide and can split easily. Though such a large mortise causes the bottom of the door to feel flimsy, it's rock-solid once the hardware is secured. Before rehanging the door, attach the end plate to the strike stile. The resistance of the end plate forces the drop bar and neoprene seal against the floor.

Auto-bottoms are especially simple to adjust. Loosen the plunger until the screw is projecting about ¼ in. out the back of the door, then close the door. The plunger's screw head will leave a dimple on the jamb, which makes it easy to locate and install the backing screw. Drill a pilot hole and drive the backing screw into the jamb, but leave it slightly proud of the jamb so that the plunger will strike the solid head of the screw. Make fine adjustments in the operation and the tightness of the seal by threading the plunger in or out of the door. I particularly like auto-bottoms that are designed to seal against the hinge side of the door first, so the neoprene never drags on the floor as the door is closing.

WEATHERSTRIPPING THE JAMB

There are five common types of weatherstripping used to seal a door against a jamb. Adjustable rigid jamb was a popular choice until early in this decade, when kerf-in weather seals like Q-Lon were introduced. Kerf-in weatherstripping is equipped with a tail fin that squeezes into a saw kerf. The kerf must be cut where the door stop meets the jamb. A third choice in weatherstripping is silicone bead, another kerf-in product. Spring bronze, a fourth weatherstripping possibility, is the most widely known weather seal, having been in use for decades. Interlock weatherstripping, though not popular, has also been used for decades. It's used so infrequently these days, though, that it's not worth mentioning here. Every type of jamb weatherstripping has advantages and disadvantages. Unfortunately there's no choice that's perfect for every opening in every home. However, knowing the differences makes it easier to choose the right product for your door.

Adjustable Rigid Jamb

Rigid-jamb weatherstripping (also known as jamb-up or adjustable jamb weatherstrip) has been widely used since the early 1950s, and it's still installed in many homes and remodels where existing jambs are not being replaced (see the bottom photo on p. 199). This weatherstripping is screwed to the jamb after the threshold is installed but before the door shoe is applied. Most weatherstripping manufacturers offer this product with either a vinyl or silicone bubble crimped into the metal retainer backing. The metal backing is screwed to the jamb, and the vinyl or silicone bubble is pressed against the door. The vinyl bubble has little spring and, unless it's installed just right on the hinge side of the door—not too close to the door but tight enough to seal—it will pinch as the door swings closed. On the strike side, the vinyl bubble also has to be installed just right—not too tight and not too loose—because it's stiff and will interfere when the latch seeks to engage the strike. Fortunately, rigid jamb weatherstripping is adjustable—the screw holes in the metal retainer backing are slotted so that the bubble can be brought closer to or pulled farther from the door.

Rigid jamb weatherstripping purchased with a silicone bubble is easier to install than that with a vinyl bubble because silicone is more pliant. Being more pliant, silicone bubbles also produce a tighter seal. Furthermore, silicone doesn't harden over time (vinyl does), it remembers its original elastic shape, and paint doesn't stick to it, at least not for very long. All these characteristics are welcome in a product that's always exposed to weather and every few years to a paintbrush. Though silicone jamb-up weatherstripping is almost twice as expensive as vinyl, it only adds about $10 per opening. Most lumberyards and large do-it-yourself centers don't stock a wide variety of weatherstripping products, so look for this material in stores that specialize in doors and door hardware.

Rigid-jamb products are also available for metal jambs. I frequently use one product that comes with two-sided tape applied to the back of the weatherstripping. Only a few predrilled slotted holes are machined into a length of this material (instead of one every 8 in. to 10 in. as in standard rigid jamb), and self-drilling sheet-metal screws rather than wood screws are included in the package.

Kerf-in Weatherstripping

In the late 1980s, Schlegel pioneered Q-Lon and revolutionized the weatherstripping industry. Q-Lon is based on a thermoset foam core wrapped with a vinyl-like skin (see the top left photo on p. 199). This S-shaped weather seal is pressed into and held secure by a thin kerf at the base of the door stop and, like the cushion seal on a refrigerator door, kerf-in foam seals the door as it closes against the stop (see the sidebar on pp. 210-211). Schlegel uses a thermoset product because it doesn't lose its elasticity and won't soften or deform. Although it's compressed by the pressure of the door, Q-Lon always springs back. Manufacturers, such as Jarrow and Amesbury Industries, also offer kerf-in weatherstripping products.

Installing Rigid Jamb Weatherstripping

Weatherstripping products are all fairly straightforward and simple to install, but a few tips might prevent minor mistakes. First of all, don't use tin snips to cut rigid-jamb weatherstripping. The aluminum bends and flattens before it cuts, which leaves an ugly scar, especially on bright finishes. Use a hacksaw, or a high-rpm circular saw with a carbide blade. A power miter saw works well, too. Cut the head first so that it fits tightly across the jamb, then shut the door and throw the dead bolt.

Press the weatherstripping against the door until the vinyl or silicone bubble just touches the surface of the door, then drive in one screw at each end. Pushing the weatherstripping too hard against the door can foul the operation of the lock, so check that the lock works smoothly before inserting more screws.

It doesn't matter which leg of the jamb is weatherstripped first, just be sure to measure from the head to the threshold. Because it almost always lands on the slope of the threshold, scribe and cut the bottom of the weatherstripping to the threshold. Be certain to crimp the aluminum over the bubble at both ends before installing the legs, so the bubble won't slide around. Like the head, press the bubble lightly against the door and drive one screw at each end and one in the middle, then check that the lock works before driving more screws. Be carful on the hinge side, too—if the weatherstripping is pushed too hard against the hinge stile, the door will pinch the bubble as it swings closed.

Kerfing Jambs for Weatherstripping

I buy my jambs precut and kerfed for weatherstripping, but sometimes I have to kerf one or two by myself. Whenever I have to, I use my table saw with a finger board. I position the rip fence so that the blade clears the face of the jamb by 1/8 in. Adding a slight angle, about 2 to 3 degrees, also helps to minimize the risk of marring the jamb, and it doesn't affect the performance of the weather-stripping. A thin-kerf blade works best, and I set the blade so that it reaches 3/8 in. past the door rabbet (see the photo below).

Router bits are also available (Norfield Tools and Supplies) that not only cut the kerf but also cut the additional 3/8-in. rabbet in a standard 1 3/4-in. rabbeted jamb. Bosch also manufactures a flush-cutting saw blade that mounts into a router and can be used for kerfing jambs.

Installing kerf-in foam weatherstripping is only hard on your fingertips, and then only if too much paint gets into the kerf before the weatherstripping is inserted. Cut the head first, square at each end, then cut the legs with a 45 degree angle at the top, so they cope over the head and form a better seal (see the top photo on the facing page). The bottoms of the legs should be cut to follow the threshold.

Fortunately, a special tool is available that makes kerfing for silicone bead much easier, and the tool will follow almost any jamb, even in a tight area. To kerf a jamb with a silicone bead tool, rest the butt of the tool 12 in. to 16 in. from the end of the jamb, then slowly plunge the head of the router into the corner between the jamb and the door stop (see the bottom photo on the facing page). The base of the tool is shaped at a 45 degree angle, which makes it easy to hold the tool in the proper position. Push lightly on the handle and cut the kerf slowly until the nose of the tool reaches the corner, then remove the tool and return to the starting point. Finish cutting the first 12 in. to 16 in. by reversing the direction of the router and slowly re-entering the previously cut kerf.

Use a tablesaw and thin-kerf blade to kerf a jamb for weatherstripping. Tipping the blade at a slight angle protects against hurting the jamb.

Kerf-in foam weatherstripping is easy to install if the kerf isn't cut too tight. It's easy to remove for painting, too.

This silicone-bead kerfing tool is nothing more than a specially molded handle equipped to hold a Bosch laminate trimmer (model 1608M). Kerfing is effortless, unless you hit a nail!

Jambs for standard kerf-in foam products are normally rabbeted an additional ⅜ in. to allow for the additional depth of the foam weatherstripping—a 1¾-in. door requires a 2⅛-in. rabbet rather than a 1¾-in. rabbet. Other sizes of kerf-in foam weatherstripping are also available, and if you encounter a jamb with a 2-in. rabbet, try a smaller size. I prefer these flexible products because they work well, they're easy to install, rarely result in callbacks or complaints, and never interfere with the operation of a door. Kerf-in weatherstripping is available in white and brown.

Silicone Bead

Q-Lon, and products that resemble it, are clearly visible between the stop and the door, which some clients find unattractive. In those cases and for retrofitting an existing jamb, silicone bead is a good alternative. This pure silicone bubble is also installed in a narrow kerf that's cut at the base of the stop on a rabbeted jamb. Because silicone bead requires no additional room (it compresses almost completely), it can be installed in existing 1¾-in. rabbeted jambs without having to rabbet the jamb any deeper. A special router base makes it easy to kerf existing openings for silicone bead in the field (see the sidebar on these two pages). Unfortunately, silicone bead isn't the most reliable weatherstripping product for this particular application, and I recommend it only for doors that are somewhat protected from the weather.

I also encounter a problem with the white silicone bead because it's translucent; seeing daylight between the door and jamb occasionally causes some homeowners concern. Fortunately, the product comes in brown, too, which isn't the only advantage of silicone bead; the material also comes in a large variety of sizes, each slightly larger than the next. I've often used three different sizes on one leg of a cross-legged jamb because each size fits perfectly inside the next larger diameter—an almost seamless piece of silicone bead can be formed to fill a gap that tapers from ⅛ in. to ⅜ in. And there's one other advantage of silicone bead weatherstripping that must be noted: This system is perfect for arched doors because the kerfing tool will follow almost any radius or ellipse, and so will the silicone bead.

Installing Cushion Weatherstripping

Most often I install cushion weatherstripping on openings that are battered by the weather, and then, ironically, I have to make all the gaps larger. It's helpful to know if the opening will be weatherstripped with cushion bronze while hanging the door because then it's easy to increase the head, strike, and hinge gaps a little. And though I don't always bevel the top of a door, I do for cushion bronze because it reduces the friction between the door and the weatherstripping; and sloping the top of the door also prevents moisture from building up and oxidizing into a green mess on the jamb and door.

I've seen spring and cushion bronze installed many different ways, but the method that has worked for me is simple. Always start with the head piece, and cut it to fit between the door rabbets. Nail the head right against the door stop. I tap each nail in until my hammer's about to hit the stop, then I use a punch to set the nail. Increased speed is not worth the risk of marring the jamb. Weatherstripping companies have special punches available that also hold the nail, but I've never needed one. Place a nail every 3 in. in the head and the jamb.

Install the legs next, but not right against the door stop. Instead, hold the leg back from the stop ¼ in. so that the leg flaps won't interfere with the head flaps. Offsetting the leg and the head increases the seal at the top of the door and reduces the angle that must be cut at the top of the leg. Cut the bottom of the leg close to the threshold, but don't cut the flap too tight or it will bind on the threshold.

To ensure a complete seal at the bottom of the door, install a corner pad at each end of the threshold. Pile corner pads are available in black and white and only cost a few cents. The bottom corners of a door are the weak spot of any weatherstripping system, and corner pads are the best insurance against light and air infiltration.

Adjusting cushion bronze weatherstripping is also easy. Use a nailset or punch and slide it inside the V-shaped flap. Bend the flap out slightly, until it contacts the door. If the flap bends too much, slide the heel of the punch over the outside of the V, and press it back down. After the door and jamb are finished, I rub paraffin wax on the cushion bronze, which allows the door to operate smoothly and stops the metal from talking.

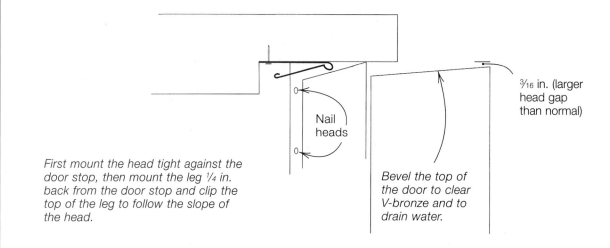

First mount the head tight against the door stop, then mount the leg ¼ in. back from the door stop and clip the top of the leg to follow the slope of the head.

Bevel the top of the door to clear V-bronze and to drain water.

³⁄₁₆ in. (larger head gap than normal)

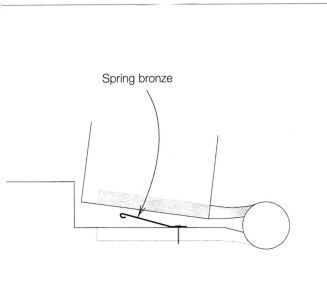

Spring bronze

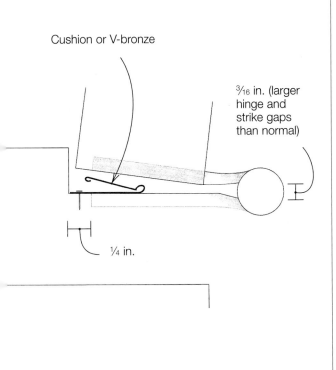

Cushion or V-bronze

3/16 in. (larger hinge and strike gaps than normal)

1/4 in.

I've used two different configurations of silicone bead. I prefer silicone corner tubeseal, available from Resource Conservation Technology. Tubeseal has a double-fin tail and secures well into a straight-cut kerf—no special cutter is required. The kerf can be cut quickly, without concern for exact depth, and sawdust is easily cleared away with a nail set. Resource Conservation also offers a router tool with a vacuum attachment, an essential component for retro-fitting doorways in furnished homes.

The second type I've tried has an arrowhead-shaped tail that requires a special arrowhead-shaped bit, somewhat like a small dovetail bit. It's more difficult to install this material than tubeseal because the depth of cut must be adjusted perfectly and if the bit and tool aren't held tightly against the jamb, the silicone bead won't be held securely by the kerf. It's also more troublesome to clean packed sawdust from the arrowhead-shaped kerf.

Spring Bronze

Spring bronze is known by several names. It's been called spring brass, cushion bronze, and V-bronze, but these designations describe different products. At a local hardware store or a large discount supplier, you're likely to find a roll of spring bronze or spring brass. Unrolled, the thin metal is attached to the rabbet of a jamb using closely spaced nails. Many jambs in older homes still have the original spring bronze installed at the time the house was built. Spring bronze works okay, though this one-piece thin metal often loses its spring and flattens out, allowing daylight and wind to penetrate a door opening.

Weather and light penetration is not a problem with cushion bronze, which is also called V-bronze. Though the material is more expensive and isn't normally available at do-it-yourself centers, cushion bronze is a vast improvement over spring bronze, and I recommend it for any opening that might pose a problem. Cushion bronze seals a doorway tightly and still allows the door and lock to function smoothly (see the right photo on p. 206). Cushion bronze is not only heavier than the thin metal sold by warehouse outlets, but it's also easy to adjust and retains tension between the jamb and door, resulting in an effective, long-term seal that won't flatten out. Cushion bronze used in conjunction with a kerf-in product can result in an almost gale-proof seal.

WEATHER-PROOFING FOR EXTERIOR DOORS

INDEX

A

Active sidelights, installing, 160, 162
Aluminum, as door component, 16-17
Arched doors, installing, 164-66
Arched jambs, building 68, 69
Astragals:
  beveled, 130, 131
  installing, 130, 131, 132, 133
  locking:
    installing, 134
    types of, 132
  overlapping, 128
  scribing for, 128-29
Auto-bottoms:
  mortising for, 207
  types of, 206-208

B

Backset, defined, 25
Back-to-back mulls, spreading, 56
Back-to-back transom head and sill, illustrated, 50
Baldwin strike plate, installing, 154
Ball-bearing hinges, when to use, 21
Ball catches, when to use, 26
Bifold doors:
  bottom pivot for, 183
  building on site, 179, 181, 183-85, 187
  cutting, 89
  designs of, 182
  exterior, hanging, 185, 186, 187
  hardware dimensions of, 180-81
  hardware for, 91
  installing, 87, 88, 89-90, 92
Bipass doors:
  installing, 92-95
  mirrored, installing, 94, 95
Bolts:
  for Dutch doors, 28
  flush:
    installing, 133, 135-36, 138-40
    when to use, 28
    slide vs. surface, 27-28
Bouvet mortise lock, illustrated, 155
Bullet catches. See Ball catches.

C

Casement latches, when to use, 26, 27
Closet doors, hinges for, 23
Colonist style door, defined, 7
Composite doors:
  edging material of, 17
  types of, 15-17
Corner hinges, when to use, 19
Cross-leg:
  adjusting, 45, 79, 81
  defined, 45
Cushion weatherstripping, installing, 212-13
Custom doors:
  layout of, 8
  when to buy, 8-9
Cutting guide, for crosscutting veneered doors, 104, 105
Cylindrical locksets, discussed, 31-32

D

Dead bolts, types of, compared, 30
Door bench, homemade, 102-103
Door holders. See Hold-opens.
Door hook:
  making, 98
  using, 97, 98
Doors:
  arched, installing, 164-66
  bifold:
    designs of, 182
    hardware dimensions of, 180-81
  composite, types of, 15-17
  double-acting, installing, 169-70, 172-75
  Dutch, installing, 166-67, 169
  exterior bifold, hanging, 185, 186, 187
  exterior, glue for, 9
  fire, 10
  hanging, 119, 122, 123, 124
  hollow-core, construction of, 3, 4
  molded-panel, construction of, 7
  paint-grade hollow-core, interior components for, 6
  pair of, hanging, 126-30, 133, 135-36, 138-42
  pocket, installing, 187, 189, 191, 193
  prehung, hinges for, 19
  solid-core hardboard, construction of, 7
  stain-grade hollow-core, interior components for, 6
  stile-and-rail, 10-14
Door shoes:
  dadoing for, 201
  installing, 203-204, 204-206
  types of, 201-202
Door skins, types of, 5
Door stand, using, 102
Door stops, when to use, 29
Door types, comparison of, 5
Double-acting doors:
  hinge layout for, 170
  installing, 169-70, 172-75
Dutch door bolts, defined, 28
Dutch doors:
  dimensions for, 168
  installing, 166-67, 169
  jamb reinforcement for, 169

E

Exterior doors, prehung, installing, 85-87
Exterior jambs:
  arched, building, 68, 69
  with full-bound sidelights, 57
  installing, 65-67
  with multiple openings, 55-57, 60-61, 63, 65-67
  with single openings, 51-55, 58-60
Exterior prehung doors, ordering, 73-75

F

Face grain, defined, 6
Fiberglass, as door component, 15-16
Fire doors:
  ratings of, 10
  when to use, 10
Flat-panel doors, 10, 13
Floor pivots, mounting, 23
Floor stops, when to use, 29
Flush-bolt locator, 143
Flush bolts:
  router templates for, 137
  types of, 28-29
  when to use, 28

G

Glue, water-resistant, for exterior doors, 9

H

Hardware, finish of, compared, 20
Hinge layout, for double-acting doors, 170
Hinges:
  adjusting, 123
  ball-bearing, when to use, 19, 21
  for closet doors, 23
  corner:
    round vs. square, 20
    when to use, 19
    determining width of, 19
    double-acting, when to use, 22
  frequency rating of, 19
  invisible, installing, 176-179
  light-duty vs. heavy-duty, 20
  locating, 101
  mortising for, 110
  piano, when to use, 22-23
  pivot, when to use, 23
  sizing, 18-19
  Soss, when to use, 22-23
  spring, when to use, 22
  wide-throw, size of, 19
Hinge stile, preparing, 107-110
Hinge swag, defined, 21
Hinge templates:
  multiple-, using, 108-109
  single-, making, 106
Hold-opens, purpose of, 29
Hollow-core doors:
  prehung, installing, 75-77, 79, 81-82
  widths of, 7

I

Interconnected locks, discussed, 33-34
Interior jambs:
  arched, building, 68
  assembling, 40
  installing, 44-50
  wall conditions for, 39
Interior openings, sizing, 39
Interior prehung doors:
  determining hand of, 72
  ordering, 71, 73
Invisible hinges. See Soss hinges.

J

Jambs:
  centering, 42, 43
  cross-sighting, 48-49
  cross-stringing, 49
  exterior:
    installing, 65-67
    with single openings, 51-55, 58-60
  interior:
    assembling, 40
    installing, 44-50

kerfing, 41
  for weatherstripping, 210-11
keying, 41
shimming, 46, 47
tweaked, straightening, 77, 78, 79
weatherstripping for, 208-209, 211, 213
Jen-Weld Door Company, mentioned, 7
Jig, boring, using, 115, 116, 117
Joinery, cope-and-stick, in stile-and-rail doors, 10-12

K

Keying jambs, on table saw, 41

L

Laminated stiles and rails, characteristics of, 13-14
Latches:
  defined, 24
  mortising for, 117, 118
  touch, when to use, 24
Lock mortiser, using, 148-49
Locks:
  finishes for, compared, 34
  interconnected, 33-34
  mortise, 34-35
  for pocket doors, 32, 33
Locksets:
  cylindrical, discussed, 31-32
  with dead bolt, 32
  installing, 113, 114
  locating, 101
  mortise, installing, 144, 145-49, 151-54, 156-57
  tubular, discussed, 30-31, 33
  types of, compared, 30-33
  See also Locks.
Lock stile, preparing, 110-18
Lock strikes:
  compared, 35
  locating, 124-25
Lock templates, making, 157

M

Medium-density fiberboard, as door component, 6
Mortise locks, discussed, 34-35
Mortise locksets:
  installing, 144, 145-49, 151-54, 156-57
  template for laying out, 150

N

Non-removable pins, purpose of, 21

P

Panels, flat vs. raised, 10, 13
Piano hinges, when to use, 22
Pivot hinges, when to use, 23
Pocket door guides, types of, 190
Pocket door kit, illustrated, 188
Pocket doors:
  bipass, 192
  installing, 187, 189, 191, 193
  lock for, 32, 33
Prefit doors:
  types of, 87
  See also Bifold doors. Bipass doors.
Prehung doors:
  exterior:
    installing, 85-87
    ordering, 73-75
  hollow-core, installing, 75-77, 79, 81-82
  installing, 75-77, 79, 81-87
  interior:
    ordering, 71, 73
    determining hand of, 72
  solid-core, installing, 82-85

R

Rails:
  bottom, width of, 10
  defined, 3
  top, width of, 10
Raised-panel doors, 10, 13
Roller latches, when to use, 26-27

S

Scribing, for doors, 97, 99-100, 127-29
Shear walls, defined, 39
Shims, making, 80
Sidelights:
  active, installing, 160, 162
  full-bound, in exterior jamb, 57
  latching, 162-63
  stationary, installing, 158-60
Sill covers:
  installing, 195-97, 204
  purpose of, 195-96
Sills:
  aluminum, advantages of, 64
  mortising, router template for, 62
  raising, 53
Slide bolts:
  vs. surface bolts, 27-28
  when to use, 27

Solid-core doors:
  latches for, 26
  prehung, installing, 82-85
Soss hinges:
  installing, 176-79
  when to use, 22
Special-order doors, types of, 7-10
Spring hinges, when to use 22
Stain-grade doors, components of, 14
Steel, as door component, 16
Stile-and-rail doors, discussed, 10-14
Stiles:
  defined, 3
  width of, 10
Stiles and rails:
  finger-jointed, construction of, 14
  laminated, using, 13-14
  veneered, using, 12-13
Surface bolts, when to use, 27

T

Template hinges, vs. non-template hinges, 21
Templates:
  for laying out mortise lockset, 150
  lock, making, 157
  multiple-hinge, using, 108-109
  router:
    for flush bolts, 137
    making, 120-21
  single-hinge, for router, 106
Thresholds, installing, 198-200
Touch latches, when to use, 24
Transoms, jambs with, 50
Trimmers:
  centering, 43-44
  plumbing, 44
  setting, 41-44
Tubular locksets, discussed, 30-31, 33

W

Warrantees, types of, 9
Weather-proofing, cutting, 196, 197
Weatherstripping:
  cushion, installing, 212-13
  kerfing jambs for, 210-11
  rigid jamb, installing, 209
  types of, for jamb, 208-209, 211, 213
Wide-throw hinges, size of, 19

PUBLISHER: Jim Childs

ACQUISITIONS EDITOR: Julie Trelstad

EDITORIAL ASSISTANTS: Carol Kasper, Karen Liljedahl

EDITOR: Carolyn Mandarano

LAYOUT ARTIST: Susan Fazekas

ILLUSTRATOR: Michael Mandarano

PHOTOGRAPHER: Gary Katz, except where noted

TYPEFACE: Stone Serif

PAPER: 70-lb. Utopia Two Matte

PRINTER: Quebecor Printing, Tennessee Book Operations